Environmental Movements around the World

Volume 1 — North America, South and Central America, Africa, and the Middle East

Environmental Movements around the World

Shades of Green in Politics and Culture

Timothy Doyle and
Sherilyn MacGregor, Editors

PRAEGER

AN IMPRINT OF ABC-CLIO, LLC
Santa Barbara, California • Denver, Colorado • Oxford, England

Library of Congress Cataloging-in-Publication Data

Environmental movements around the world : shades of green in politics and culture / [edited by] Timothy Doyle and Sherilyn MacGregor.
p. cm.
Includes index.
Volume 1—North America, South and Central America ; Volume 2—Europe, Asia and Oceania
ISBN 978-0-313-39353-2 (hardback) — ISBN 978-0-313-39354-9 (ebook)
1. Environmentalism. 2. Environmental sciences. 3. Environmental policy.
I. Doyle, Timothy, 1960–
GE195.E584 2014
363.7—dc23 2013020879

ISBN: 978-0-313-39353-2
EISBN: 978-0-313-39354-9

18 17 16 15 14 1 2 3 4 5

This book is also available on the World Wide Web as an eBook.
Visit www.abc-clio.com for details.

Praeger
An Imprint of ABC-CLIO, LLC

ABC-CLIO, LLC
130 Cremona Drive, P.O. Box 1911
Santa Barbara, California 93116-1911

This book is printed on acid-free paper ∞

Manufactured in the United States of America

Contents

Part III: Africa and the Middle East

Acknowledgments

This book is designed for students, researchers, activists, and others who are interested in the many shades and shapes of environmentalism around the world. Coming from different countries and cultures, and presenting a range of perspectives, the contributors share a common commitment to placing environmental movements in the context of larger struggles for eco-political change, globally and locally. We dedicate this book to them and to all those engaged in these necessary struggles.

We owe tremendous thanks to Dr. Adela Alfonsi, whom we relied on heavily as assistant editor, and without whose expert editorial assistance, diplomacy, sound judgment, patience, and kindness, this project simply would not have been possible. Without her willingness and ability to work closely with authors whose first language is not English in order to improve their chapters, we would have a very different kind of book.

We would also like to thank the team at Praeger/ABC-CLIO who believed in this project from the start and who saw it through to production. Emily Birch, Beth Ptalis, Anthony Chiffolo, and Valavil Lydia Shinoj (Apex CoVantage) deserve extra special mention. Our colleagues at Keele University and the University of Adelaide provided support in a number of ways, including making it possible for Sherilyn to spend time in Adelaide to providing funding for Adela's work. We would also like to acknowledge that significant parts of this two volume title draw upon the research of a much larger three-year project entitled "Building an Indian Ocean Region DP120101166," which is funded by the Australian Research Council (ARC) Discovery Projects Scheme for funding in 2012/2015.

Finally, we thank our family members, especially Fiona Lawrence and Simon Pardoe, for their understanding and encouragement throughout the project.

Chapter 1

Green Tide: "Worlding" Environmental Movements

Timothy Doyle and Sherilyn MacGregor

> At midnight the crowd returned to the traveller. "Time's up," it said, speaking in its 1001 voices, not in perfect unison. "Answer the riddle. Explain us. What's our definition? What makes us real?" He replied, "But whatever definition I gave would exclude half of you. It's the lack of definition that unites you."[1]

Environmental issues—toxic pollution, deforestation, changing climates—are global and local at the same time. So too are the movements of people that champion them. In every part of the world, when people gather together to address environmental problems and to search collectively for solutions, we say they are engaged in "environmentalism." If their struggles go on for years, gain momentum and infamous leaders, experience hard knocks and sometimes victories, we call this an "environmental movement." Since their first emergence in the late 1960s in the global North, environmental movements have been an unstoppable tsunami, a giant green tide of people flowing across all global oceanic geopolitical systems—the Atlantic, the Indian, and the Pacific oceans—at first touching the shores of all regions and countries through their engagement with traditional politics, and then moving inland to challenge the very fabric of cultures, religions, and other belief structures. That environmental movements operate in a context of economic and cultural globalization is certain. What is highly debatable is that there is a global environmental movement. The central argument of this collection is that there is no single environmental movement: it is a phenomenon of many voices and many shades of green. We still need country-level analyses, because even when environmental movements in different countries are subject to the same global forces, they do not all react in the same way.[2] More importantly, resisting simplistic homogenization is necessary to improve understanding and foster political solidarity among environmentalists from around the world.

Because it is inherently plural, we favor the term *environmentalisms.* It is a term that defies simple definition. Let us make it clear at the outset that it is not for us, as Northern scholars, to define environmentalism or to determine who is a *true* environmentalist, or who is not, by drawing up ideal types or devising Western-centric taxonomies. Our aim is to open up the field to a range of definitions and voices, many of which have not been included before now in a conversation that has a more than 40-year history. For the purposes of this collection, we use the words *environment* and *green* interchangeably. People in different cultures and countries utilize either/both of these words (and others) to describe the selection of issues raised by human relationships to the natural world. What is important is that the experiences described in the following pages are written by and about environmentalists whose grounding in diverse locales makes it possible to present environmental movements within their cultural and political contexts. We have left it to them to explore the current shapes and shades of environmentalism in a way that fits their own understanding and responds to their own local concerns. Pursuing this position of respecting local definitions, we do not split hairs about what constitutes a social movement, an environmental movement, a green movement and so on. We start from the basic stance of accepting how people within a given context define the issues that define their movements. We refer to this as an *emic* approach: that is, we attempt to see the movements through the eyes of the people themselves. As a consequence, we allow space for different framings. For example, animal rights issues traditionally have been seen as lying outside the environmentalist frame, whereas in the Australian case, the rights of animals have always occupied a central place in the movement. We do not advocate a particular political definition of environmental movements: they are simply very broad coalitions of individuals, networks, groups, organizations, and parties oriented around the *environment* in all its interpretations and diverse guises. They may be organized around single issues or pursue comprehensive platforms. In some cases, these movements have collective radical intention to disrupt the status quo within their societies; in others, they are strategically engaged with the dominant polities of their states in order to effect change from within. What they share is a desire to confront and change the structures and practices that cause harm to natural environments and to the individuals and communities whose lives and livelihoods depend on them. It is this common ground that we explore in this collection, while at the same time presenting a challenge to those who believe that there is one unified environmental movement.

Our Approach: "Worlding" Environmentalism

In introducing this collection, we want to make explicit some important points that will enable readers to understand the approach and the inspiration behind the project. It is an ambitious project that we think is unique within the field of environmental politics; it has been two years in the making and is the product of transnational collaboration.

First, a word about us, the editors. We are Northern scholars with an academic base in the United Kingdom, but "from" the opposite ends of the planet: Tim Doyle is from Australia; Sherilyn MacGregor is Canadian. We teach environmental politics together at Keele University in the very middle of England. This project was inspired by a desire to demonstrate and explore the diversity of environmentalisms that often gets lost in a field that is ever-obsessed with the global, the transnational, and the climate. It also allows us to draw on our many years of research and writing on environmentalism. Tim Doyle has written extensively on environmental movements in the majority world of the global South, on postcolonial people's movements against extractive industries and megaprojects, largely in the Indian Ocean Region.[3] Sherilyn MacGregor has written on women's environmental activism, with a focus on movements in the affluent global North, and has an interest in both the theoretical concept and practical enactment of ecological citizenship.[4] Rather than collaborating on a monograph on environmentalism around the world, however, we decided that this had to be an edited collection. Too often in these days of academic neoliberalism, where individual self-interest and preservation drives research, edited collections are seen as a "poor cousin" of monographs. But there was no question for us that a collection not only made sense but was consistent with the substance and spirit of the project. There are already many books on green politics by Anglo-Europeans and Americans. Why should we in the minority world get to define what environmentalism is by speaking on behalf of people living in places we've never lived in and who speak languages we don't speak?

Through our international networks, we set out to find some of the very best writers on green movements within their national contexts. We approached these scholars/writers/activists (we do not separate ideas and actions as so often happens in the global North) and asked them to write about the current state of environmentalism in their countries and cultures. We provided a broad remit, which was to give readers a sense of the historical, social, political, and cultural context in which particular examples of environmentalism are located and to describe the ideas that inform the movements and their practices. However, we did not ask authors to conform to a universal blueprint; rather we encouraged them to pursue their own connections to and understanding of environmentalism. Some have focused on distinct campaigns or movements, and others have given broad overviews. Including the voices of local writers, immersed within their own national and regional cultures, has resulted in a richer and more grounded work than could possibly have been achieved by a monograph by us.

Producing a collection of chapters written by scholars from around the world is consistent with a "worlding" approach that has been defined eloquently by international relations (IR) scholars Ole Waever and Arlene B. Tickner. Their edited collection, *International Relations Scholarship Around the World: Worlding beyond the West,* endeavors to ameliorate the domination of the discipline by Americans by including perspectives from scholars working at the margins of the geocultural order who have a different interpretation of core IR concepts (such as sovereignty, security, and power) than those working at the center. They write that "an emphasis on worlding brings to light the question of the situatedness of knowledge and experience in relation to the dominant readings of world globalization, and simultaneously constitutes a call for reconnecting to particular anchorings of practices in social situations, for 'worldliness.'"[5] Their book contains country case studies written by IR scholars from parts of the world that have hitherto been absent from the disciplinary conversation. Their approach and their excellent book have inspired us to ask questions about the available body of research on environmental movements. How much of the field has been defined from a Western or Northern perspective? Although there are notable exceptions, the bulk of the literature on environmentalisms takes the American framing to be normal and imposes it on the rest of the world. Environmentalisms that have taken root in authoritarian regimes or that operate within societies shaped by the legacy of colonialism do not fit nicely within this frame, yet we rarely have an opportunity to find out what alternative framings are possible. Following the Waever and Tickner example, this collection aims to decenter the field of environmental movement scholarship by giving voice to authors from 24 countries and making our voices, and the voices of North Americans and Western Europeans, just a few among many—in fact we are nicely outnumbered.

Worlding environmentalism is not so simple in practice; perhaps this is why there are few if any other collections of this kind. The world in the 21st century is dominated by English language and academic publishing, in peer-reviewed journals for example, and operates on an assumption of English language proficiency. Most work in the social sciences comes from the minority world, and there are observable geopolitical power relations within academia that mirror the relations of domination and dependency between core and periphery.[6] Getting published as a nonnative English speaker from a university or organization with scarce resources (e.g., limited access to online journals) and few network connections is incredibly challenging. This means that their valuable observations and analyses will be lost unless they receive extraordinary support from those privileged enough to give it. One of us recently heard the editor of a top academic journal dedicated to the academic study of geopolitics say that his weekends are too important to him to have to spend time fixing the grammar of nonnative English-speaking authors. Unlike this editor, it is because we are committed to worlding that our project required us to work with the authors to edit their chapters into academic English, and in some cases to translate chapters written in other languages. The process of translation is not just about replacing one word for another; it requires an understanding of the working conditions of scholars from different parts of the world and having a profound respect for their working conditions. Early in the project, we were fortunate to find, in Adela Alfonsi, a translator who is not only a skilled editor but someone capable of understanding the complex social science issues that pertain to environmental issues and movements. Her involvement in producing these chapters was indispensable. As a consequence, we invited her to collaborate with us in writing the final chapter of Volume two.

Structure of the Collection and This Volume

This collection presents case studies of the distinctive kinds of environmental politics—including ideologies, tactical repertoires, strategies, and broad political and campaign approaches—that are espoused and practiced within specific countries. It presents state-of-the-art research on the ideas and actions of environmental movements around the world. It explores the myriad ways in which people in different countries and regions have interpreted environmentalism, from a range of political, scientific, economic, cultural, philosophical, and religious viewpoints. The aim is to examine what people do when acting environmentally: the collective ideologies, mythologies, and cosmologies of the different environmentalisms.

There are two volumes in the collection, each incorporating countries from the global North and the global South. When we use the terms *global South* and *global North,* we are not imagining a clear separation in a geographical sense, neatly splitting the northern and southern hemispheres. Instead, the global North is portrayed as the affluent world, the world of the economically and politically powerful minority. It is sometimes called "the minority world" to indicate that although the richest and most powerful, it is but a fraction of the world's population. The global South, on the other hand, depicts the lives of the majority world, the less-affluent realities of the vast majority of the people on the planet. These divides often, but not always, flow through the formal political boundaries of nation-states and regions. It is possible to be part of the global South within a nation normally considered to be part of the global North— for example, some indigenous peoples and other less-affluent minorities who live on the peripheries of shining global megacities. The opposite is true: powerful and wealthy minorities (such as those working in the financial sector) living in countries usually earmarked as being part of the global South, actually inhabit the transnational worlds of the global North.

The chapters in Volume one are organized into three global geopolitical regions: South and Central America; North America; and Africa and the Middle East. These chapters are rich and engaging contributions from outstanding scholars/activists working in these parts of the world.

Part One: South and Central American

We start in South and Central America showcasing studies from five countries: Chile, Brazil, Venezuela, Argentina, and Mexico. We begin with the chapter by Claudia Sepúlveda and Pablo Villarroel, which gives an overview of Chilean environmentalism. Going back to the time of the most recent dictatorship, the writers highlight the particularities and sociopolitical and cultural factors underpinning Chilean environmentalism. They elucidate two distinctive features: first, Pinochet's legacy and the long exposure to neoliberal economic programs, uncritically supported by all subsequent Chilean governments; and second, the tensions between international and internal environmental demands. Importantly, among the main strands of Chilean environmentalism are traditions of what might be called new spiritualisms: philosophical approaches that influence key environmental leaders and Green parties. Indeed the role of spirituality and religion is a theme that emerges in quite a few of the cases throughout the two volumes, and one we will revisit in the second volume. While this spiritualist tradition underpins much of Chile's environmentalism, the institutional framework is one of accommodation between elites, as they develop a consensus style of politics after emerging from the dictatorship, which is both neoliberal and nonconflictual. For example, the environmental impact assessment (EIA) system is oriented toward the interests of business, which takes on assessment and monitoring roles that should belong to the state and squeezes out public involvement in policy-making and discussion.

The EIA process in Chile fails in a series of cases, detailed in the chapter, demonstrating its inefficiency and antidemocratic nature, thereby raising doubts about its legitimacy. This lack of legitimacy leads to institutional breakdown. As a result of the Valdivian ecological disaster in 2009, there is a "before and after" effect in Chilean environmentalism. There has been some reform of environmental processes, but the authors argue that these still have not been democratized. Since the end of the 2000s, international conservation nongovernmental organizations (NGOs) have also made their own mark, advancing their agendas and ways of working, but managing to displace the Chilean NGOs and reinforcing the preexisting technical approaches. Hence, a process of moderation within the environmental discourse has been observable, and some depoliticization, but there are signs of some important changes being made, driven by popular demand for greater democratization.

The second South American case study is taken from Brazil. Benito Cao's fascinating chapter takes us to the struggle over the future of the Brazilian Amazon, that vast river ecosystem that still defines one of the last remaining tracts of tropical rainforest on the planet. So often, this area has been called "The Lungs of the World." Cao explores the link between local, national, and international dimensions of environmental activism in Brazil and the problems to which they respond, by focusing on the movement to prevent the construction of the Belo Monte dam. In the first part of the chapter, he explores the history of environmentalism in the Amazon. He writes about the Portuguese "civilizing" discourse as applied to the Amazon and its inhabitants and how this narrative of exploration and exploitation is reflected in current government policies that succeed in marginalizing indigenous communities. The overview of environmentalism in the Amazon includes the role of Chico Mendes in forging an alliance between green concerns and the economic and social needs of rubber tappers. Before his murder in 1988, Mendes successfully internationalized the issue via the "boomerang effect": lobbying on a world stage came back to put pressure on the authorities in Brazil. However, there is an important local context to be considered. Upon emerging from dictatorship, the new Brazilian Constitution sets aside the Amazon as a national treasure and recognizes the cultural and territorial rights of indigenous peoples, thus providing an avenue for judicial contestation that has been applied, both successfully and unsuccessfully, until current times.

The second part of Cao's chapter examines resistance to the Belo Monte Dam. The "glocalization" of the issue in relation to the protection of the Amazon has its own problematics: there is access to new funds, and the involvement of foreign, more resourced NGOs, but this also means a more professionalized and bureaucratized approach that has on the one hand strengthened environmentalism, and on the other hand, depoliticized it. The market-friendly approaches applied by some NGOs (for example, consumer awareness campaigns around the derivation of some consumer products or certification processes) deprive the indigenous peoples of their agency. Cao explores the indigenous Kayapó resistance movement, along with their capacity to make themselves the center of a diverse network of alliances in the region. This activism gives indigenous politics a regional dimension, apart from enlisting other local social movements and appealing to international celebrities, such as James Cameron. The Kayapó have taken their denunciations to international agencies including the UN Human Rights Council. There are threats of violence, and from time to time, direct action. This directly challenges the state's commitment to economic developmental policies, and raises fears about national sovereignty over the Amazon. It allows the government and other supporters of the dam to argue that foreigners are agents of Western imperialism and that resisters of the dam are actually an obstacle to the economic progress of the nation. Economic development and national security concerns exercise the mind of the state in the case of the Belo Monte dam. The issue has divided the government, with developmentalism having the upper hand. On top of that, there is also the problem of endemic political corruption and the impunity with which laws can be and are flouted, and the incidence of abuses never brought to trial. In the struggle between developmentalism and protection of indigenous rights, Cao sees the latter as the only real means of securing the protection of the Amazon.

In her chapter, María Pilar García-Guadilla describes how Venezuela's government espouses a New Left "socialism of the 21st century" while applying a logic of extractivism by extracting hydrocarbons, mainly oil to develop Venezuela. Extractivism is in contradiction with the stated aims of social justice and poverty reduction; the government's developmental policies in fact cause gross violations of human and environmental rights. García-Guadilla examines the government's revolutionary discourse, and the national and regional development plans, to demonstrate their contradictions and the conflicts and damage that they cause. She also looks at the mobilization of a network of social and environmental movements in response to government-aided problems. But what are these socio-environmental movements? How many have been co-opted by the government because of resources offered or because they agree with the ideology that the revolution must come before the environment, and that the former will protect the latter. Poignantly, she argues that the dilemma for New Left governments in Latin America is how to eliminate the dependency on oil and the export of fossil fuels without which it is difficult to imagine the successful emergence of transformative and emancipatory social movements.

The final South American chapter is, in fact, a comparative case study of three countries, one of which, Brazil, has already been addressed by Cao. The other two countries, Argentina and Mexico, also provide useful material for exploring similarities and differences between movements for environmental justice in Central and South America. Taking us up on our invitation to approach environmentalism broadly, Mauricio Berger decided that this comparative form suited him better than a single case study approach. We were happy to accommodate the move away from a single-country case study, not only to reflect cultural differences in environmental movement experience, but also the experience and expertise of this particular writer. In this vein, then, Berger explores the creative responses to environmental injustices in three countries that share a situation where the state itself is accused as the perpetrator. In each case, the state is to blame for environmental injustice because it is captive to capitalism, violates human rights, and, as a result, has lost its legitimacy. Through these three case studies, Berger argues that environmental justice networks provide deliberative spaces and alternative mechanisms for organizing civil society in the fight for environmental and social justice. These horizontal, open networks are nonhierarchical, informal, autonomous nodes with no central command or formal structure. Their functions include making problems visible, developing alternative mechanisms for community monitoring and reporting, and providing education. Some of the activities Berger discusses include "interdisciplinary clinics," "collective counter-cartographies," "popular and critical epidemiology," "popular EIAs," and "community surveillance." He considers these to be acts of institutional and political creativity, "creating other kinds of political institutions" that "unfold the functions that correspond to the state but without the state."

Part Two: North America

We next move northward to Canada and the United States. It becomes immediately apparent, as we jump from the majority world to the minority world, that the environmental movement phenomena being described are indeed "worlds apart" even though we are still in the Americas. Nowhere else is the global North–global South divide more accurately and starkly expressed than here.

Laurie Adkin's comprehensive study of Canadian environmentalisms presents a movement defined by two opposing approaches, with differences in issue-focus and regional identity.

It is worth noting that the importance of region is felt throughout this collection, both at the larger geopolitical and continental level as well as in recognizing the regions and subregions within particular nation-states. In this vein, Adkin discusses how specific intrastate geographical localities reflect different political economies and different issues, and this allows her to "map" Canadian green movement experiences along ideological lines as well.

Despite exploring the various differences, Adkin also paints a picture of the dominant form of environmental tradition that is currently entrenched in Canada and may be on the rise. This raises a crucial point: although the 24 case studies in our collection prompt a celebration of diversity, this is not to discount the fact that certain voices and narratives can dominate at any given time, thereby silencing other voices. The cacophony of movement noises is still subject, like any form of human collective endeavor, to the disciplining sounds of the most powerful. Environmental movements are not simply about "goodies" (greens) versus "baddies" (polluting and extractionist corporations). But while environmental movements as a rule work to contest the dominant ideologies of large, transnational corporations, there are ideological struggles that take place within movements themselves. Green movements more often than not reflect the political realm of the societies in which they operate. So it is futile to imagine that environmentalists are all on the side of emancipation and pursuing development pathways that oppositionally challenge existing hierarchies and inequalities. Oftentimes, as Adkin explains, movements—whether willingly or under coercion—speak power-to-power, seeking environmental change within the infrastructure and architecture of the dominant social frame.

Social movements—and, for the purposes of this collection, environmental movements are just one type of social movement—can be places of fierce contestation between diverse ideologies. In the Canadian case, Adkin describes the dominant ideology driving the current movement as being supportive of market-driven, neoliberal environmental policies, not because environmentalists like them, but because they do not feel they have much choice. In this story, green NGOs often appear co-opted, depoliticized, and excessively genteel. But she ends her analysis by providing a song of hope: over the last two decades there has been a shift in the dominant discourse, as Canadian environmental movement organizations (EMOs) develop deeper connections with more radical social movements in the South, themselves oriented more toward social and environmental justice. Indigenous discourses are also beginning to play a greater role in debates. Therefore, we have a dichotomy here, not just based on region or subregion but based on an ideological divide: some groups move toward a more reformist and institutional position whilst others take a more radical stand. Adkin concludes by arguing that it is currently unclear as to whether things are moving in a different direction or whether the environmental movement in Canada stands divided. This contest between those who prefer to play politics-as-usual or business-as-usual and those who seek political change on a fundamental, systemic level is a key characteristic of green movement experience, in the affluent global North in particular.

From Canada we go south again to the United States of America, still a superpower in global affairs despite recent shift in global geopolitics. In his chapter, Robert Brulle draws on decades of research in the field to provide a broad sweep of the history of American environmental movements, with a detailed description of their frames and discourses. Brulle is interested in looking at a multiplicity of different movements within one country (very much in line with our plural sense of environmentalisms) rather than seeing different networks within one unified movement. He paints a picture of U.S. environmentalisms in decline, its political clout gradually eroding since the 1990s. Despite these movements' long and rich histories, organizational complexity, and public support, he believes that the retreat is due to the rise of neoconservatism and the decline of liberalism. There can be no doubt in the U.S. context that the countermobilization against environmentalism by the right has been extremely successful. The Wise Use Movement, for example, first emerged in the United States in the 1980s, funded by neocons, populists, and large extractive industries, based on the strategic belief that the only way to "beat" the green movement was to mobilize on the ground in their own social movement. Whether the Wise Use Movement is a movement at all remains an open question. But many of the think tanks and the NGOs that rose under its rubric are, in reality, *corporate front groups* capturing the rhetoric and narratives of green movements, and then using these to strike against them. Cleverly, the large extractive companies poured money into these front groups, infiltrating and dissolving the adversarial messages of environmentalism.

Brulle concludes his chapter with the rise of professional environmental advocacy and the limitations of mainstream environmental discourses. He argues that American environmental movements and their ascendant discourses have been depoliticized, and the general public, as a result, has been marginalized by the increasingly technically expert nature of environmentalisms. No alternative vision is being produced and, as a consequence, most green organizations focus on curing symptoms rather than providing a meaningful, democratic platform for elaborating a genuinely sustainable society. In the very end of his contribution, Brulle proposes a theory of transition, of "directed social change," which involves reconnecting with the idea of a just and sustainable society, with environmental groups necessarily creating their own institutions of media transmission. In short, he wants expert knowledge to be "at service of and accountable to nonprofessional publics."

Nowhere else is this "expert capture" of environmentalism more apparent than in the climate discourse in the global North. Regardless of its obvious validity of the issue in an "essentialist" ecological sense (i.e., human-induced climate change exists and is a huge global problem), it must be recognized that it is driven by Western, *white* science and other elite knowledge systems, and this certainly explains the removal of elite environmental activist engagement with the more public, grassroots realms and networks of the movement. In this manner, unfortunately, green politics in the United States has gone the way of so many social movements within the affluent minority world: it has been cordoned off into elite pockets of power, and whether we can call the dominant cartel of activists a social movement at all becomes a point of debate. On the positive side, the environment justice movement, which first emerged in the late 1980s and early 1990s, still provides a key challenge to these dominant voices, arguing for a more equitable planet that recognizes that after 40 years of environmentalism in the global North, there is still an overarching unwillingness in the North to address issues of fair access to limited resources and to own up to gross disparities of wealth within their own cultures. In this way, the U.S. environmental justice movement argues that black communities, and communities of color, are more likely to live next to a toxic waste dump than white communities.[7] And even more recently, the rights of indigenous peoples in the United States are now included in these stories of environmental justice. This takes us back nicely to our earlier discussion of the global South existing both among and within nation states, predominantly defined by affluent world experiences. The global South often lives within the global North and vice versa, and powerful elites comprise global North networks within countries most usually depicted as inhabiting the peripheral realms of the global South.

Part Three: Africa and the Middle East

Next, we move from the South-in-the-North to the South-in-the-South Part three of this volume consists of two chapters on countries in Africa followed by two from the Middle East. To begin, Janet Muthoni Muthuki focuses on a key characteristic of green movements in the global South: the central role of women. The Kenyan Green Belt Movement (GBM) is famous for its community-based tree planting program, its role in empowering women, and because of the leadership of the late Wangari Maathai, the first African woman to win the Nobel Peace Prize (2004). Here it is presented as a movement that is a form of African eco-development that places women, class, citizenship, and the protection of the environment in the same frame. Women are the daily managers of the environment, and Muthuki argues that in traditional Kenyan culture, men and women have always had differentiated roles in environmental management and that these roles made management sustainable. Communal land management and holistic views of the environment were part of the traditional system. More recent land demarcation systems have destroyed traditional social structure, thanks to colonialism, and more damage was subsequently done by the Kenyan state and its application of modernizing discourses. With traditional social structures gone, African patriarchy is aggravating the damage done by Western patriarchy. The GBM is not just about tree planting but about women's empowerment with its focus on civic education, advocacy, teaching, good governance, and seeking democratic reform.

Moving from the Indian Ocean side of Africa to the Atlantic side, John Agbonifo takes us to the equally celebrated case of popular environmental struggles in Nigeria. He provides an excellent taxonomy of environmentalisms that can be found in that rapidly developing, oil-rich East African country. Like Muthuki, Agbonifo argues that African environmental ethics and practices are very different from Western environmental ethics and practices. Furthermore, Western secular and economic reductionisms should be resisted, regardless of whether they are badged as "green" or otherwise. In contrast to what we often see in the West, African environmental ethics are more holistic and metaphysical, and he takes some care to explain and illustrate what he means by this. Finally, he draws on the case of the Ogoni people of the Niger Delta to demonstrate that their collective action frames draw on their religious and cultural repertoires. Their demands are expressed this way, and their leaders behave not according to elitist political opportunity theories drafted somewhere in Western Europe or North America, but according to concepts of "divine commission" or purpose.

Once upon a time, being green was seen as some kind of luxury pursuit, a kind of purely "post-materialist" political program pursued only by white, Western, middle-class people. And no doubt, in some parts of the affluent world, this had been the case. The African stories presented in these two chapters, however, show us that this assumption can no longer hold. Indeed, we must now see that the least affluent people on the planet can be green, but in their own way, their own shade. It is not just a middle-class, white, minority-world phenomenon when we accept that environmentalisms are founded on very different interpretations of environmental relationships, needs, harms, and wishes. What then does environmentalism look like in a war zone? Are there environmental movements in such places? Mohamed Salih, in a brave contribution to this collection, looks at environmentalism in Iraq, beginning his piece with an overview of the types of environmental destruction that were caused by the Iraq War.

Salih tells a story of the Marsh Arabs to demonstrate that single-issue environmental movements do not exist in Iraq. Instead, they have multiple "hybrid identities." For the Marsh Arabs, protection of the environment was also a case of their own physical survival and need for self-determination in the face of an authoritarian regime. As such, it is also a social justice issue. Hybrids are identities that span the boundaries of two or more social movements, for example by bridging feminist, labor, peace, or antiracist struggles. The Marsh Arabs' political framing brings together issues of self-determination, social justice, environmental security, and human rights (targeted because of their ethnicity and political affiliation). In Salih's view, hybridity is particularly likely to be experienced under authoritarian regimes, because the lack of democracy means governments can engage in violent, abusive and repressive acts more readily. Post–Saddam

Hussein, environmental NGOs are as much concerned with political, social, and economic issues as they are with the traditional notions of environment as "nature." The Western response to the Iraq war is also hybrid, he argues. Because of concerns about climate change, environmental movements are usually antiwar, in alliance with multiple other identities or single-issue movements. But these issues are all understood to be interconnected. Western environmentalism is also global/local—responding to the global problem of climate change, but articulating its responses via domestic political processes. Salih concludes his argument that single-issue global environmentalism is at a crossroads, and must join forces with other social movements, such as those tackling the ravages of poverty and war, in order to address the concerns of the majority of the world's population.

These global and local crossroads are a perfect segue to a neighboring Middle Eastern country: Israel. In the final chapter of this volume, Alon Tal also speaks of Israeli environmentalism at a crossroads. Drawing on years of first-hand experience of environmental advocacy and green party politics, he begins by acknowledging that Israel has all the elements that should make it a spectacular success story in environmental movement terms. It has a strong cultural background supportive of environmentalism (shorn up by religious and nationalistic roots), public awareness, a strong environmental movement, and strong legal and political institutional settings. Yet, contrary to these favorable background conditions, Tal outlines severe environmental problems within the movement and with Israel's ability to resolve critical environmental problems.

In this vein, Tal identifies three key issues that stand in the way of positive environmental change in Israel: population density, a culture of consumption, and the green movement as a "political orphan." In the first case, there is the immigration of Jewish people to Israel, and more importantly, a very high birth rate in the ultra-orthodox resident Jewish community, bolstered by pro-natalist government policies. The acute religious and political sensitivities around these issues make conversations about them very tricky, so there is a tendency to avoid. Secondly, there is the issue of the culture of conspicuous consumption. This he attributes to what he calls "a crisis of values." Finally, on the issue of environmentalism as a "political orphan," Tal laments that the political structure of the Israeli state favors small parties and a political dynamic whereby pressure groups can pressure these small parties for concessions to secure narrow sectoral objectives. Unfortunately, at the national level, because of the prevalence of these smaller interests, where it would be possible to portray environmentalism "as a public good for all," there has never been an opportunity to gain political support by a major political party. In the end, it remains not enough of an issue to gain electoral traction. The political context is of small parties in coalition within an overarching frame of traditional security as *the* imperative.

* * * * * *

These stories from across the planet that inspire this collection are a celebration of environmentalism's diversity. They are stories that capture the profoundly different ideas that inform environmentalists, their actions and repertoires, and the ways in which they engage with their political and social systems. Although there is no uniformity, we would not want to say that there are no shared ideas and experiences. What we have found in the process of putting this work together is that similarities are more likely to be found in regions and neighboring countries than across the globe as a whole. The simple question as to *what environmentalism means* is up for grabs in each chapter. More stories and more voices can be found in Volume two. We hope readers will follow the green tide as it moves from shore to shore across this vast and vulnerable globe.

Notes

1. Rushdie, *Midnight's Children.*
2. Doherty and Doyle, "Beyond Borders," 701.
3. Doyle, *Environmental Movements in the Majority and Minority Worlds.*
4. MacGregor, *Beyond Mothering Earth.*
5. Waever and Tickner, "Geocultural Epistemologies," 10.
6. Alatas, "Academic Dependency and the Global Division of Labour in the Social Sciences."
7. Bullard, *Dumping in Dixie.*

Bibliography

Alatas, S. F. "Academic Dependency and the Global Division of Labour in the Social Sciences." *Current Sociology* 51, no. 6 (2003): 599–613.

Bullard, R. *Dumping in Dixie: Race, Class and Environmental Quality,* Third Edition. Boulder: Westview Press, 2000.

Doherty, B., and T. Doyle. "Beyond Borders: Transnational Politics, Social Movements and Modern Environmentalisms." *Environmental Politics* 15, no. 2 (2006): 697–712.

Doyle, T. *Environmental Movements in Minority and Majority Worlds: A Global Perspective.* Piscataway, NJ: Rutgers University Press, 2004.
MacGregor, S. *Beyond Mothering Earth: Ecological Citizenship and the Politics of Care.* Vancouver: University of British Columbia Press, 2006.
Rushdie, S. *Midnight's Children,* quoted in The Riddle of Midnight, an Antelope Production for Channel Four, 1988, screened on SBS (Australia) on "Masterpiece," February 1989.
Waever, O., and A. B. Tickner. "Geocultural Epistemologies." In *International Relations Scholarship around the World: Worlding beyond the West,* edited by A. B. Tickner and O. Waever, 1–31. London and New York: Routledge, 2009.

PART I

South and Central America

Chapter 2

From a Scientifically Oriented Environmentalism to a Grassroots Call for Democracy

Claudia Sepúlveda and Pablo Villarroel

Introduction

This chapter analyzes the distinctive features of Chile's environmentalism from a historical perspective. Our aim is to reflect on the sociopolitical and cultural factors that contribute to explain the emergence and evolution of the Chilean environmental movement along with its particularities, underlying tensions, and contemporary trends. Our focus is placed, on the one hand, on the constitution of environmental organizations, their discourses and agendas, as well as in the ways in which they have been conditioned or shaped by political and institutional contexts. On the other hand, we pay attention to the influence that environmental movements have exerted over the design, performance, and reform of the country's environmental framework.

Political and institutional contexts have played a key role in shaping Chile's environmental movement. In particular, 17 years of dictatorship (1973–1990) and its most salient legacy—that is, the world's first and perhaps most extended neoliberal program[1]—have been determinant in the profile and strategies of environmentally concerned actors. We argue here that the fundamental effect of both historical events—Pinochet's dictatorship and the implantation of the Chilean neoliberal program—has been to moderate the political component of environmental discourses and strategies, restricting social participation and favoring the scientifically and technically oriented approaches that have characterized some of the most influential strands of the Chilean environmental movement.

This scientific orientation can be traced to the first wave of conservationist organizations in the 1960s. These organizations were mainly focused on the protection of animals and ecosystems, and have been described as properly ecocentric.[2] Although they did not necessarily oppose more politicized strands, their emphasis was on technical rather than political visions, and in the scientific analysis and solution of problems rather than in sociopolitical change.[3] This scientific orientation was deepened during dictatorship, when the political scenario favored a tactical connection between environmental NGOs and the academic community. Such an alliance reinforced the technical orientation that Chile's environmentalism inherited from the conservationist pioneers.[4]

Two other distinctive features characterize Chile's environmentalism. The first one concerns the enduring hegemony of the neoliberal framework installed as Pinochet's legacy and deepened under the democratic period. In particular, the country's nascent environmental institutions in the early 1990s were designed in consistency with such neoliberal frame, imprinting a dynamics of social exclusion in environmental decision making that persists until today.[5] The uncritical backing that the neoliberal program received from the four democratic governments of the center-left *Concertación* (1990–2010) blocked alternative agendas and triggered a grassroots demand for extended participation in environmental decision making.[6]

A second feature refers to the tension between internal and external environmental demands. External demands related to prospective commercial agreements (i.e., NAFTA) and international trade played a determining role in the settlement of Chile's first environmental frame in 1994.[7] These demands were consistent with the priority that the *Concertación* gave to economic growth and the need to comply with environmental standards of international markets in order to guarantee economic stability and competitiveness.[8] Thus, despite an already consolidated environmental movement, Chile's first environmental frame was designed not in response to its demands but to those of international markets, aligned with national elites. This pattern changed from the 2000s as an effect of the maturing of an internal demand for the democratization of environmental decision making. Its most visible expression occurred in 2004 when a local movement emerged in Valdivia, in the south of Chile, in response to a major ecological disaster.[9] The sociopolitical resonance of this local mobilization triggered a major revision of the country's environmental frame, finally forcing its total reform by the end of 2009.[10] Contemporary environmentalism is dominated by the consolidation of this internal demand for environmental democracy, expressed in an increasing number of grassroots movements and local ecological struggles.

In what follows we expand the revision of these characteristic features of Chile's environmentalism concluding with a deepened reflection of its contemporary trends.

The Main Strands of Chilean Environmentalism

Chile's environmental movement has been described as constituted by three differentiated strands that took form since the 1960s: conservationism, environmentalism, and ecologism.[11] Each of these strands is considered foundational and their distinctive ideologies and agendas can be traced through to the present.[12]

Conservationism is focused on the protection of natural habitats and species. This strand has important links to ecological sciences and its arguments tend to be framed in scientific or ethical terms. Its perspective is ecocentric and rather neutral politically. In Chile, this approach characterized the first generation of environmental organizations born in the 1960s and has continued to be a major component of the environmental movement.

Environmentalism is concerned with environmental problems such as pollution and resource degradation. Its general perspective can be described as anthropocentric. Politically, it tends to support the reform of social and economic structures rather than radical change. In Chile this approach has been adopted by NGOs of academic orientation born during dictatorship and focused on the solution of environmental problems. Right-wing think tanks created during the 1990s are also situated within this tradition.

Ecologism is also characterized as anthropocentric, although it makes an explicit connection between environmental degradation and the threats posed to life supporting systems. In both cases the final causes are attributed to the capitalist order and the human domination of nature while solutions are conceived in the form of radical transformation. In Chile this strand is represented by several NGOs, some connected to political platforms and mostly created under dictatorship.

To these strands Carlos Aldunate[13] has added a fourth one: the tradition of the *new spiritualisms* as a foundational component of the Chilean environmental movement. In brief, new spiritualisms are philosophically grounded approaches that interpret the environmental crisis in terms of the paradigms that underpin human-nature relations. Most spiritualist traditions prioritize changes of mentalities and practices rather than of social structures, albeit oriented to broader processes of collective transformation. As we will see, this strand has played a fundamental role in the political expression of Chile's environmentalism. On the one hand, through its connection to some prominent environmental leaders, and on the other, through its involvement in the foundation of the two green parties that have so far existed in the country.[14]

Historical Stages of the Environmental Movement

Chile's environmental movement can be organized in four phases: the *foundational stage* (1960–1973), the *dictatorship stage* (1973–1990), the stage of the *set in force of environmental institutions* (1990–2004), and the stage of the *breakdown and reform of environmental institutions* (2005–2012).

During the *foundational stage,* the environmental movement was dominated by a small group of conservationist NGOs, rather disconnected from political contingency. Although not yet visible, the new spiritualist tradition also appeared during this phase.[15] The development of these emergent organizations was violently interrupted by Pinochet's coup d'état in 1973.

Despite the politically adverse context and the total neglect of the environment during the military regime,[16] environmental NGOs diversified in the *dictatorship stage.* Moreover, the profile of Chile's environmentalism took a definitive form during this period. New environmental NGOs committed to social justice and democracy developed novel adaptations of social-ecological and political-ecological approaches.[17] In turn, new spiritualist approaches adhered to discourses of sociopolitical transformation and developed the first experiences of environmental activism. However, it was the consolidation of the environmentalist tradition—that is, of scientifically oriented, anthropocentric perspectives—that allowed environmental NGOs to be acknowledged as key actors in debates around the first ever environmental frame during the late 1980s and early 1990s.

The stage of the *set in force of environmental institutions* began with the return of democracy in 1990 and is marked by the approval of the environmental law in 1994. During this stage, all founding traditions lived a period of expansion due to the opening of spaces for debate and participation. While ecologist and new spiritualist traditions fostered the first-ever green political platforms, technical cadres from conservationist and environmentalist NGOs got involved in the design of nascent environmental policies. The work of these environmental actors increased in complexity and visibility along with the explosion of environmental conflicts. In turn, an entirely new generation of environmental organizations emerged, including grassroots actors and international conservationist organizations that first landed in the country

The stage of *breakdown and reform of environmental institutions* began in the mid-2000s, coinciding with a crisis of legitimacy of the country's environmental frame. During this period historic NGOs lost salience partly due to a reduction in international funding and the prevalence of the neoliberal consensus, which marginalized critical stances. As a result, environmental NGOs prioritized lobby activities, became disconnected from grassroots actors, and shifted to more technical discourses. Simultaneously, an inorganic network of local movements emerged in connection with the dynamics of conflictivity that has characterized environmental impact assessment processes. The fundamental demand of these local actors is a deepened participation in environmental decision making, particularly with respect to megaprojects with negative effects over communities and territories. In sum, while environmental NGOs have lost their resources, vitality, and political grip, a still not fully shaped local demand for environmental democracy has become the main trend of Chile's contemporary environmentalism.

The Birth of Chile's Environmentalism: Conservationist Foundations in the 1960s and 1970s

A first wave of local environmental NGOs and associations emerged in the mid-1960s. Rather disconnected from the experiences on social transformation that dominated a then politicized Chilean society, this nascent environmentalism was concerned with the depletion of natural resources and oriented toward wildlife conservation.[18]

The first environmental NGO created in Chile, the Comité Nacional pro Defensa de la Flora y Fauna, CODEFF (National Committee for the Defence of the Flora and Fauna), settled the base for the conservationist lineage of the country's environmental movement. Created in 1968 as a membership association by the lawyer Godofredo Stutzin, CODEFF has remained one of the most influential environmental organizations. Stutzin's fundamental ideas, considered as properly ecocentric,[19] pointed to an ethic of respect and compassion with other forms of life, which he materialized in the proposal of legal instruments for the protection of animals and nature more broadly.[20] CODEFF joined the International Union for the Conservation of Nature (IUCN) in 1968, showing an early connection with the international environmental movement.

CODEFF's vision reflected a broader conservationist sensibility that was beginning to resonate in other actors, particularly within intellectuals and politicians. President Eduardo Frei Montalva (1964–1970), for example, was especially concerned with soil erosion and the overexploitation of native forests.[21] Accordingly, he inaugurated one of the largest reforestation programs ever implemented in Chile, along with studies and inventories of native forests.[22] The justification of such initiatives, however, was not environmental but economical. As President Frei himself explained, his vision was that Chile could conquer a prominent position in the international market of paper and pulp through 450,000 hectares of forest plantations.[23] This forest policy would produce critical effects after four decades of implementation, expressed, for example, in the ecological disaster that occurred in Valdivia in 2004, which we will further analyze.

Although less visible, spiritual sensibilities related to national as well as international thinkers also fed into the discourses and philosophical foundations of Chile's environmentalism during the 1960s and 1970s. Spiritualist groups shared a common disbelief of both capitalist and socialist ideologies, proposing an individual transformation as the basis for any meaningful social change. Two of its figures were of particular relevance: the psychiatrist Lola Hoffmann and the acknowledged biologist Humberto Maturana. Maturana founded what later became known as the Santiago School of Cognition in collaboration with Francisco Varela, developing a line of thought based on the notion of *autopoiesis*. These ideas inspired notable ethical reflections on the ecological crisis. However, the influence of the spiritualist tradition within the environmental movement would only become evident from the 1980s.[24]

In 1982, a group of 500 individuals founded the Planetary Initiative for the World We Chose (Iniciativa Planetaria para el Mundo que Elegimos), connected to its Canadian version.[25] This was the first organic form of the spiritualist tradition, exerting important influence within intellectual circles.[26] Some of the most salient environmental NGOs created during the 1980s and 1990s, such as Defensores del Bosque Chileno (Defenders of the Chilean Forest) and Casa de la Paz (The Peace House), were born from the work of some of its members. Maturana himself would take public leadership on debates about the destruction of native forests in the early 1990s.[27] And more importantly, organizations pertaining to the spiritualist lineage were directly linked to the creation of Chile's—and Latin America's—first green parties in the late 1980s and 1990s.

The expansion of this first generation of environmentally concerned organizations was drastically interrupted by the 1973 military coup, condemning them to a marginal position for more than a decade along with other expressions of organized citizenry.

Diversification under Dictatorship: Ecologism and the "Scientific Turn" in the 1980s and 1990s

A more diverse generation of environmentalism emerged under Pinochet's dictatorship, comprised of three different branches: *social ecology, political ecology*, and *environmentalism*. The first two founded what we here call the tradition of *ecologism*. Overall, their most distinctive feature was a more explicit concern with social justice and democracy as well as a clearer political stance. The tradition of *environmentalism*, in turn, was born in connection to academic circles. As Clapp[28] has noted, such a connection was a strategy to gain legitimacy and survive under particularly adverse political conditions.

Indeed, during the 1970s and 1980s official discourses were dominated by hostile visions on the environment. Nature was conceived as a set of resources that had nothing else than economic values.[29] The main representatives of the Corporación Chilena de la Madera, CORMA (The Chilean Timber Association), for example, proposed that native forests should be massively substituted by exotic plantations for they were "disordered" and dominated by "old" and "overmature" individuals, and thus, were economically valueless.[30]

Within this context, an important group of academic NGOs began to develop the *social ecologist* tradition. Most were founded under the umbrella of the Catholic Church, which became a refuge for scholars expelled from public universities. The Centro de Estudios y Promoción de Alternativas Urbanas, CEPAUR (Center for Studies and Promotion of Urban Alternatives), is noteworthy for the work of the economist Manfred Max-Neef, who inaugurated ecological economics in Chile. Max-Neef elaborated his famous human-scale development theory for which he was awarded the Alternative Nobel Prize in 1983. Max-Neef's work was steeped in the *ecologist* and *spiritualist* traditions, whose members named him the first ever candidate of the environmental movement in the 1993 presidential campaign where he obtained 5 percent of the votes. Such a result marked the peak of the political expression of Chile's environmentalism.

Inspired by CEPAUR's work, a salient group of academically oriented NGOs developed applied experiences on ecological economics through socially appropriate technologies for sustainable development in urban and rural communities.[31] These included solar panels, adobe building, organic agriculture, and efficient water systems.

The *political ecological* branch within the ecologist tradition also appeared during the dictatorship.[32] This branch was comprised of a group of NGOs that emerged in the late 1980s, ideologically positioned within perspectives of social and political ecology.[33] Its most notable representatives were the Instituto de Ecología Política, IEP (Institute of Political Ecology), founded in 1987 by Manuel Baquedano, and the Red Nacional de Acción Ecológica, RENACE (National Network for Ecological Action), founded in 1989 under the umbrella of the IEP by Sara Larraín.

IEP's work was oriented to develop a political program to rearticulate the Left around a green platform.[34] RENACE, in turn, worked to establish a base of social support for such a platform, coordinating a group of 80 grassroots organizations.[35] As spokesperson for this network Larraín became a well-known environmental leader. Almost a decade after, in 1997, she founded the Programa Chile Sustentable (Sustainable Chile Program), a renewed effort to consolidate a green political platform whose culmination was her candidacy for the presidency in the 1999 campaign. Larraín competed against Tomás Hirsh, a candidate from the Humanist Party that represented the green agenda of the new spiritualist tradition. She obtained 0.44 percent of the votes, far from Max-Neef's 5 percent in 1993, while Hirsh obtained 0.5 percent.

The *environmentalist* tradition, in turn, first emerged through the Centro Interdisciplinario de Desarrollo Urbano y Regional, CIDU, (Interdisciplinary Center for Urban and Regional Development), created by the architect Guillermo Geisse as an academic unit within the Universidad Católica de Chile (Chilean Catholic University). From this same space Geisse founded the Centro de Investigación y Planificación del Medio Ambiente, CIPMA (Center for Environmental Research and Planning), in 1979. CIPMA's work marked a turning point regarding the social legitimacy of the environmental agenda.[36] Its main achievement was to bring into the environmental field top scientific figures, including the national scientists' awards, the physiologist Héctor Croxatto and the physicist Igor Saavedra, and the prominent philosopher Juan Gómez Millas. During the First Scientific Meeting for the Environment organized by CIPMA in 1983, this NGO burst onto the scene as a new environmental actor. With no evident connections to the conservationist or spiritualist strands, CIPMA inaugurated the tradition that became known as *environmentalism.*[37]

CIPMA organized seven subsequent scientific meetings every three years that became the most important space for the articulation of the still emergent environmental movement.[38] Positioned under this academic umbrella, CIPMA's scientific encounters adopted a pragmatist tone that privileged the dialogue between different perspectives, including those of the business sector. Moreover, CIPMA received the economic support of Chilean as well as international corporations, contributing to disseminate the notion of sustainable development that helped to validate the environmental discourse of productive actors.[39] Overall, the most important effect of CIPMA's work was to consolidate the scientific approach of Chile's environmental movement.

The mid and late 1980s was also the period of consolidation of the spiritualist tradition, which took a definitive political expression when a group of organizations launched the ecopacifist network Futuro Verde (Green Future). With colorful street protests and solid work with local communities, Futuro Verde was the first activist environmental organization. Two of its most salient campaigns were related to Santiago's air pollution and the defense of the Chungará Lake inside a national park over 4,000 meters in the Chilean Andes that was threatened by the extraction of its water for agriculture.[40] By the end of 1988 *Futuro Verde* turned into the Los Verdes (The Greens) political party, the first legal green party of Latin America.[41] The Los Verdes' platform was oriented to promote "a new mentality" and to defend the "quality of life" through fighting the oppression, social injustice, and discrimination of "the prevailing system."[42] The party was explicitly framed as nonpolitical.[43]

In 1989 Los Verdes took part in the nascent *Concertación* with two deputy candidates for the first democratic election after dictatorship. Since they obtained less than 5 percent of the votes, and no elected representative, they formed the Humanist-Green Alliance with the Humanist Party, also from the governing *Concertación.* The Humanist-Green Alliance stepped out of the *Concertación* in 1993, arguing that the governing coalition had abandoned its founding principles and uncritically embraced the neoliberal agenda.[44] In 2002 a group of nonhistorical leaders unsuccessfully attempted to reorganize the Greens.[45]

Chile's 1994 Neoliberal Environmental Frame

Once Patricio Aylwin was elected as the first postdictatorial president, in 1989, environmentally concerned actors turned their efforts to the challenge of establishing the country's first environmental framework. In 1988 a group of scientists, under the umbrella of the Comisión Nacional de Investigación Científica y Tecnológica, CONICYT (National Commission for Scientific and Technological Research), prepared a first draft of an environmental law inspired by the agreements reached on the Second Scientific Meeting on the Environment organized by CIPMA in 1986.[46] Under CIPMA's umbrella a selected group of experts, which included academics as well as lawyers and business representatives, prepared their own recommendations for the coming law.[47] Both texts were core inputs for the environmental bill sent by Aylwin to the Congress in 1992 and approved in April 1994.[48]

The historically elitist style of politics that had prevailed along Chile's history, and was reactivated during the transition to democracy, also dominated the establishment of the country's first environmental law: its contents were negotiated between political and economic elites with the exclusion of environmental actors that had been determinant in the recovery of democracy.[49] As described by David Carruthers,[50] under the *Concertación* Chile's political elitism took the specific shape of the *democracia de los acuerdos* (democracy of agreements)—a style of politics aimed to avoid conflict, prioritize a pragmatic efficiency through the exclusion of interferences, and focus on direct negotiations between a handful of selected leaders from each relevant party. Within this context, scientifically oriented NGOs assumed the role of experts, collaborating with the governing elites who took in their hands the definition of the institutional building that would mark Chile's environmental history for the next decade and a half.

The basic legal design agreed upon by these elites implied the reinforcement of the neoliberal program.[51] In particular, the bill elaborated by Aylwin's administration created a coordinating agency with the task of promoting the inclusion of environmental issues in all sectorial ministries and public policies, instead of a centralized ministry. In turn, the bill proposed a gradualist approach through a basic institutional frame to be improved over time. The resulting design was, in Carruthers's[52] terms, "remarkably simple" compared to those in place in most countries, including Latin American ones.

Basically, the Chilean model was comprised of a general environmental "framework law," as it was called, a decentralized and coordinating entity called the Comisión Nacional del Medio Ambiente, CONAMA (National Commission for the Environment), and regional environmental commissions, Comisiones Regionales del Medio Ambiente, COREMAs. While the CONAMA was in charge of designing the general policies and promoting new legal standards in collaboration with the responsible ministries, the COREMAs were comprised of appointees of national authorities whose main task concerned decision making process within the Sistema de Evaluación de Impacto Ambiental, SEIA, (Environmental Impact Assessment System). In the absence of any meaningful legal definition regarding sectorial standards or land-use planning, the SEIA became in practice the main policy instrument of the new institutional design.[53] Moreover, as Eduardo Silva[54] has noted, hardcore issues, such as environmental standards for productive sectors, were entirely left to regulatory bodies to be designed away from public scrutiny. Furthermore, whenever developed, such environmental standards were exclusively referred to end-of-the-pipe emissions but never to industrial processes. Furthermore, critically polluting sectors, such as the pulp industry, which constitutes the most important export product after mining, lack any specific standards for their dangerous emissions.

The revision of the bill in parliament was marked by a fundamental event: the prospective incorporation of Chile as a member of an expanded version of the North American Free Trade Agreement (NAFTA).[55] The U.S. Congress had conditioned Chile's incorporation to the existence of minimum environmental legislation, and in particular to some form of environmental impact assessment proceeding.[56] As explained by David Tecklin, Carl Bauer and Manuel Prieto, business representatives, whose opinion was determinant for obtaining right-wing support for the bill in the Congress, have acknowledged that although the environmental law was neither considered important nor a priority, NAFTA was both.[57]

In turn, the timeline imposed by Chile's negotiations with the NAFTA were critical to incline the government's option toward a low-profile organization, such as CONAMA's coordinating model.[58] Furthermore, such a model was explicitly backed by the World Bank, which granted the government a $17 million loan to implement the environmental reform.[59] Despite the World Bank's acknowledgment that the coordinating model was a "risky experiment," such backing was intended to reinforce the principles of state subsidiarity and decentralization that had been at the base of Chile's neoliberal state restructuring during the 1970s and 1980s.[60]

It may be concluded, then, that Chile's first environmental law was designed and approved in response to the interests of pragmatic leaders who saw in the expansion of trade agreements an opportunity to secure the sustained economic growth that they considered fundamental for a smooth transition to democracy. In other terms, it was an external demand, imposed by legal environmental standards required by external markets, that constituted the main driver behind the approval of the environmental law in 1994 by Congress.[61]

It should not be surprising, thus, that the new law was consistent with the prevailing neoliberal consensus supported by the governing elites. Indeed, the environmental frame was oriented to enable rather than to regulate markets, reflecting the neoliberal principle of minimizing the intervention of the state in favor of the free operation of markets and private actors.[62] This market-enabling approach was expressed in very specific aspects, being most evident in the design and functioning of the SEIA.

In brief, the SEIA was conceived as a "one-stop-shop" (*ventanilla única*) for environmental permits, that is, as a unique legal revision of new investments, minimizing the transaction costs associated with the new environmental regulations.[63] Another neoliberal expression of the SEIA was that it left in the hands of investors key functions related to the assessment and monitoring of environmental impacts that otherwise should correspond to critical tasks of public agencies.[64] In particular, investors were deemed responsible for preparing the ecological baselines and the impact assessments reports for their projects, as well as the monitoring of reports of the eventual impacts generated by their own projects.[65] Moreover, and given the total absence of land-planning mechanisms, proponents had the freedom to choose any location they considered appropriate for their operations. In the context of an extremely market-friendly institutional setting many companies have abused such freedom by choosing to locate their projects, for example, inside or near protected areas. Despite social opposition, some of these locations have ended up producing devastating effects such as the ones that occurred in Valdivia, as we will see further.[66]

The neoliberal conception underlying the environmental frame was also expressed in its limited spaces of public involvement. Although the SEIA's new participatory mechanisms were unprecedented for the Chilean context, their overt limitations fuelled an increasing frustration amongst citizens.[67] In particular, public involvement was restricted to written observations prepared by the affected parties within a window of 60 days beginning from the moment when the project was officially submitted to the SEIA. Thus, participation formally began once fundamental decisions related to the location and technological design of the projects had already been taken.[68] Moreover, any meaningful changes introduced to the projects under assessment were excluded from public participation if the 60 days window was due.[69]

The Gradual Emergence of an Internal Environmental Demand

From the late 1970s, but especially from the early 1990s, a series of environmental conflicts exploded in the public scene. Their escalating intensity can only be understood in connection to the ideological dispute they expressed. Still dominant, extremely neoliberalized conceptions over nature were publicly opposed by NGOs, citizens, and even by the Catholic Church.[70] The most emblematic of these conflicts were related to large-scale projects that represented threats to native forests and local communities. The dispute around the Astillas de Chiloé (Chiloé Woodchips) project, in the Chiloé Island, exploded in 1978, anticipating the intensity of social resistance to environmentally destructive investments. It was followed by long-lasting emblematic struggles, such as those triggered by the Pangue dam in the Bío-Bío River in 1990, the Terranova woodchips and substitution project in Valdivia in 1990, the Trillium project in the lenga forests of Tierra del Fuego, in the Magallanes Region in 1993, and the Ralco dam also in the Bío-Bío basin in 1994.[71] For the first time local communities emerged as environmentally concerned actors, although strongly supported by NGOs.

Indeed, the more politicized expressions of the environmental movement, in particular the ecologist strand represented by IEP and RENACE, became increasingly involved in these local disputes. New NGOs even emerged to support communities involved in disputes, such as the Observatorio Latinoamericano de Conflictos Ambientales, OLCA (Latin American Watchdog of Environmental Conflicts), created under IEP's umbrella in 1992. OLCA grew from a little team to an independent organization whose work for the defense of environmental rights has been increasingly acknowledged.

The burst of local environmental disputes also prompted the constitution of the *bancada verde* (green congress-bench) in the early 1990s, a group of deputies from Left- and Right-wing political parties, self-described as environmentally inclined, who played a key role in some paradigmatic environmental controversies during the 1990s.[72]

However, it were conservationist NGOs such as CODEFF and Defensores del Bosque Chileno, founded in 1994, who stood as the most salient groups in the public debate, confirming the weight of this conservationist lineage. Indeed, the fact that the single most relevant environmental issue of public debate in the 1990s was precisely the controversy around the conservation or substitution of native forests[73] reflects the dominance of conservationists' discourses. An important factor that contributed to intensify such debate was the landing in Chile of the American conservationist Douglas Tompkins. In 1991 Tompkins launched the creation of the 300,000 hectares Pumalín Park, triggering fierce opposition from economic and political elites concerned by the geopolitical and territorial implications of this mega-protected area. Although he never intervened openly, Tompkins became an important supporter of many conservationist NGOs during the 1990s, partially explaining their stability and success. The influence of the conservationist lineage is also reflected in the fact that the biologist Adriana Hoffman, founder of Defensores del Bosque Chileno and one of the most well-known environmental leaders, was named in the highest public environmental position as director of the CONAMA in 2000, although she stayed for less than two years.

Once the 1994 environmental frame was established, communities expected to influence SEIA's final decisions through its newly created participatory mechanisms. Shortly, however, they confirmed their total lack of power.[74] Even those communities that chose to play the game, occupying every participatory space provided, concluded that public involvement in the SEIA was designed not to involve citizens in environmental decisions, but rather to neutralize their opposition against decisions that had already been taken by investors and authorities.[75]

This technocratic approach to environmental participation gradually triggered the rise of an increasing conflictivity around decision making in the SEIA.[76] By the end of the 1990s, after only half a decade of operation, the SEIA was already exposed to an escalating crisis of legitimacy. In particular, during Eduardo Frei Ruiz-Tagle's administration (1994–2000), a series of high-profile environmental disputes occurred within the SEIA, involving strategic megaprojects. These included the U.S. Trillium forest project in Tierra del Fuego, Magallanes, the Ralco dam of the then Chilean energy holding, ENDESA, in the Bío-Bío river basin, and also the Chilean forest and pulp giant CELCO-ARAUCO's pulp-mill in Valdivia, all submitted to the SEIA between 1995 and 1996.[77] These three conflicts also had in common the fact that the projects had been rejected by the COREMAs' technical committees but later approved by the SEIA after President Frei had publicly backed each of them.[78]

Sometimes overt—as with President Frei—but mostly invisible, the systematic political intervention rendered null any meaningful public participation within the SEIA. Even though the SEIA showed a rejection rate of around 18 percent until 2003,[79] in the case of the most politicized projects—such as the Ralco dam or CELCO-ARAUCO's pulp-mill—political interventions have clearly favored the industry.[80] Such a pattern reflected President Frei's statement that under his administration "no investment would be stopped for environmental considerations."[81] In practice this meant that the COREMAs' efforts should be placed in the environmental improvement of projects subject to environmental assessment, despite the type and magnitude of their deficiencies, rather than their rejection. Frei's tenet became known, following Claudia Sepúlveda,[82] as the *doctrina Frei de impactos ambientales consumados* (Frei doctrine of *fait accompli* environmental impacts), leading to the absurdity that the Trillium project, for example, was approved with more than 100 "conditions" instead of being rejected for its substantive weaknesses.

In the absence of land-planning procedures that could counterbalance the case-by-case environmental decision making by the SEIA, an increased unrest related to the environmental assessment of investments gradually led to a crisis of legitimacy of the whole environmental building.[83] The culmination of such a crisis occurred as the effect of a local ecological struggle that exploded in Valdivia, in the south of Chile, which we address in the following section.

During this period of settlement of the environmental frame, environmentalist NGOs such as CIPMA and, overall, academics and experts, consolidated their influential position within decision making circles. In turn, conservationists and ecologist organizations, such as CODEFF, *Defensores del Bosque Chileno,* IEP, and RENACE, played a more critical role by exposing the neoliberal orientation of Chile's environmental policies to public scrutiny. However, the challenges related to this important role led them to partially sacrifice their connection to grassroots actors, as expressed, for example, in the 0.44 percent obtained by Sara Larraín in the 1999 presidential election.

In turn, new NGOs emerged. The Agrupación de Ingenieros Forestales por el Bosque Nativo (Association of Forest Engineers for the Native Forests) was created in 1993, opening an important space for the involvement of forest engineers in public policies for the sustainable management of native forests. In 1997 the Terram Foundation was created, orienting its work to environmental policy-making, environmental participation, forest conservation, and sustainable salmoniculture. Finally, *Ecoceanos* (Ecoceans) was born in 1998, focusing on marine conservation in connection to the livelihood of coastal communities. All these NGOs can be defined as politically critical, more close to *ecologist* than to *environmentalist* traditions.

International organizations also began to land in Chile during this phase. The first one was Greenpeace, which opened its office in Santiago in 1994. It was followed by the World Wildlife Fund for Nature created in 2002 in Valdivia, and by the Chile programs of Oceana in 2003 and of The Nature Conservancy in 2004.

Institutional Breakdown: The Valdivian Disaster and the 2009 Environmental Reform

An increasing number of conflicts connected to the questioning of environmental assessment processes within the SEIA had been accumulating since the early 1990s. Altogether they represented a structural crisis of legitimacy of environmental decision making that became evident by the end of 2004, when a protected wetland located upstream the city of Valdivia, in southern Chile, was heavily polluted by the effluents of a new pulp mill owned by the Chilean giant CELCO-ARAUCO.[84] The Valdivian disaster triggered one of the most emblematic ecological struggles that ever occurred in the country, also forcing a deep revision of the environmental frame that led to its total reform by the end of 2009.[85]

In February 2004 CELCO-ARAUCO's pulp mill began to operate, discharging its liquid wastes to the Río Cruces, 30 kilometers upstream the Carlos Anwandter Sanctuary, a nationally and internationally protected area of more than 4,700 hectares that was the habitat of 119 species of birds, amphibians, and aquatic mammals.[86] In only a few months the most important reproductive population of the endangered black-necked swans dropped from near 9,000 in January 2004 to less than 200 in January 2005.[87] Other species and also economic activities were drastically affected.[88]

The location of CELCO-ARAUCO's discharges to the Río Cruces had been objected to during the environmental assessment of the pulp mill in the mid-1990s, when it became the first project to be submitted to the SEIA in the Lakes Region, not only by Valdivian actors but also by the COREMA's technical committee.[89] Both pointed to the damage that the industrial wastes could cause to the protected wetland. However, then-president Frei's public backing for the project forced the COREMA to approve it, although making it conditional to two options: the addition of a tertiary or chemical treatment before discharging the liquid wastes to the Río Cruces, or their disposal into the ocean. The company attempted the latter option but confronted fierce resistance by the coastal community of Mehuín.[90] Finally, CELCO-ARAUCO added the tertiary treatment and the project was approved in October 1998.[91]

The massive death of the swans, confirmed in October 2004, sparked an unprecedented local movement that called itself *Acción por los Cisnes* (Action for the Swans), demanding the detention of CELCO-ARAUCO's operation until its connection to the disaster was dismissed.[92] *Acción por los Cisnes* focused its work on reviewing the conditions under which the plant had been approved by the SEIA. The movement made public an amazing list of unlawful acts committed by CELCO-ARAUCO, including the construction of a plant with a productive capacity at least 68 percent higher than what was authorized.[93] Simultaneously, *Acción por los Cisnes* revealed the institutional failures that had ended up producing the wetland's ecological collapse.

In response to the Valdivian movement's demands, the CONAMA hired a scientific study to the Universidad Austral de Chile (UACH), based in Valdivia, in order to determine the causes of the disaster. After five months of fieldwork and experiments, UACH's team concluded that CELCO-ARAUCO's discharges were enough to explain the changes that had occurred in the ecosystem.[94] One of UACH's main findings was that CELCO-ARAUCO had been discharging great amounts of non-authorized chemicals to the Río Cruces, which the company had not reported despite its presence being confirmed by official measures from independent agencies.[95] Although UACH's report pointed to these compounds as causally related to the disaster, environmental authorities granted the company an ad hoc permit allowing them to continue their discharge without the environmental assessment legally required.[96]

Since 2006 environmental agencies have ordered administrative measures aimed at restoring the wetland. However, CELCO-ARAUCO has continued to discharge its wastes to the Río Cruces and at the time of writing this chapter, the ecosystem had not yet recovered.[97] Based on the UACH's scientific report the Chilean state placed a lawsuit against CELCO-ARAUCO through the Consejo de Defensa del Estado (Council for the Defence of the State) in April 2005. The final resolution is still pending.

Despite the apparent failure of the Valdivian movement, it succeeded in exposing the voids that had led to the institutional production of the Río Cruces disaster. As a result, it generated a historical consensus about the need for a major reform of the 1994 environmental frame.[98] In particular, representatives of the most important business associations, who had resisted previous attempts to substantially modify the environmental law, agreed that what they called "the scandal of the swans" represented a "before and after" in the country's environmental setting.[99] Simultaneously, a critical report on Chile's environmental performance was published by the Organization for Economic Co-operation and Development (OECD) in October of 2005. The report raised a series of recommendations for Chile to become a full OECD member, emphasizing the need to broaden spaces for public participation and to introduce land-use planning procedures.[100]

The environmental reform also became a core topic in the 2005 presidential campaign. The *Concertación's* candidate, Michelle Bachelet (2006–2010), dedicated a whole section of her platform to detail the measures for expanding the timeline of the SEIA's participation mechanisms, supporting communities involved in environmental assessment processes, and introducing citizen involvement in land-use planning.[101]

In 2007 Michelle Bachelet named the first ever minister for the environment with the task of reforming the country's environmental setting. The corresponding bill was sent to the Congress on July 2008, proposing the creation of three new entities: a ministry for the environment, an oversight environmental agency (Superintendencia del Medio Ambiente) and an environmental assessment service (Servicio de Evaluación Ambiental, SEA) that would replace the infamous SEIA.[102] With the creation of the SEA, the COREMAs would be replaced by a technical regional entity that was expected to put an end to the intervention in environmental decision making by political authorities. The bill did not consider, however, any meaningful advance with regard to citizen participation, except an extended timeline for public involvement in case meaningful changes were made to projects under assessment.[103]

The chamber of deputies introduced important amendments to the proposed bill, including mandatory participation in the newly created strategic environmental assessment of public policies.[104] The senate discussion, in turn, generated unusual attention, expressed in more than 1,000 amendments to the bill. However, Bachelet's government conducted a closed negotiation to guarantee the support of right-wing senators, blocking the debate and imposing agreements that were never discussed in the Congress.[105] These agreements resulted, for example, in the elimination of the mandatory participation that had been introduced in the lower chamber and the reactivation of the broadly questioned COREMAs.[106] In addition, Bachelet's government agreed to pass a new law proposed by right-wing actors for the creation of environmental tribunals that would replace ordinary courts in all environmental matters, with authority to, for example, reverse extreme fines ordered by the newly created environmental oversight agency.[107] The reform of the environmental frame was approved in November 2009, under the terms negotiated with right-wing senators by Bachelet's administration. In turn, the environmental tribunals were approved in January 2012.

The 2009 reform represented an important advance in terms of raising the priority of the environmental policy, strengthening the power of the environmental authority, and unifying procedures for the supervision and sanctioning of law violations.[108] However, it largely failed in fulfilling the demands for the democratization of environmental decision making that had been raised by local environmental actors for more than a decade.[109] Thus, in the end, it reinforced the authoritarian and technocratic visions that underpin the escalating socio-environmental unrest that characterizes Chile's contemporary scene.[110]

Recent Trends: Grassroots Movements and the Call for Democracy

The arguments and demands of the Valdivian movement received unprecedented support at a national level, resonating with dozens of similar conflicts that had been occurring within the SEIA since its inception.[111] Overall, the Valdivian mobilization expressed the historical consolidation of an internal demand for the democratization of environmental institutions and, in particular, for meaningful public involvement and legitimate decisions. That is, decisions not subject to the political intervention through which national authorities had been imposing their own priorities, even against the SEIA's technical organs.

Since the end of the 2000s, the local arena has been colonized by conservationist international NGOs, not always articulated to the agendas of national actors. Rather, these foreign organizations have brought their own programs and have focused part of their work on the lobby needed to make room for them. The consequence has been a displacement of Chilean conservationist NGOs and, once more, the reinforcement of technical approaches, this time disconnected from native perspectives.

In turn, NGOs have tended to adopt more scientifically inclined and even technocratic and elitist styles of work. This is expressed, for example, in the priority given to lobby activities and the involvement in formal process for the design of environmental policy. In these spaces, many NGO leaders and professionals have acted as spokespersons of the environmental movement despite their organic disconnection from the grassroots actors that they are supposed to be representing. In the end, thus, most NGOs have come to play, more or less, the role of experts, albeit critical, abandoning what was left of political projects of social transformation.[112]

Such a role corresponds in great part with a sharp reduction in international funds, which has affected the scope of environmental NGOs and their programs. While many NGOs currently operate with minimum staff, others have searched for available resources, including their transformation in consultants. In addition, the neoliberal consensus that prevailed during the 20 years of the *Concertación's* administration dramatically reduced the spaces for public dissent. As a result radical stances were marginalized while NGOs were forced to moderate their discourses if willing to be considered legitimate counterparts by the governing elites. The overall result has been a depoliticization of Chile's environmental NGOs.

This depoliticizing trend has begun to show some signs of reversal during Sebastián Piñera's administration (2010–2014), under which social mobilization—centrally including environmental demands—has burst into the public sphere. In response, *ecologist* NGOs such as the Programa Chile Sustentable as well as new actors such as the Observatorio Ciudadano (Citizens Watchdog), founded in Temuco in 2004,[113] have initiated new strategic alliances with local communities and social actors involved in environmental conflicts, who have become the protagonists of Chile's contemporary environmentalism. In particular, Programa Chile Sustentable has made important efforts aimed to build a national network of grassroots organizations actors around water issues, which gas gained increasing political relevance.

Indeed, citizens show signs of being more and more committed to environmental issues. Several surveys have systematically demonstrated the priority given to environmental problems—such as water pollution and the protection of native ecosystems—by the Chilean population.[114] Two studies conducted in 1999 and 2005 in the Lakes Region concluded that the majority of its urban inhabitants (75 percent) are willing to sacrifice part of the economic growth and their own incomes in order to favor environmental protection.[115] In turn, a survey conducted in 2009 by TNC, that represents the 98 percent of the country's voters, urban and rural, concluded that 58 percent agrees that environmental protection should be a priority even if it implies a loss of jobs or investments in the short term.[116] Environmentally concerned citizens have also demonstrated their ability to exert acute political and institutional effects, such as those generated by the Valdivian mobilization.

In the long run, then, what has persisted are citizens' demands for the democratization of environmental decision making even after the 2009 environmental reform was set in force. Such a trend is demonstrated by the historically massive mobilizations that occurred in 2011 in Santiago and other cities against what was considered an undemocratic approval of the Hidroaysén hydroelectric project, in the southern Aysén region.[117] This controversial investment, by the country's two biggest electricity-generating companies—ENDESA and Colbún—could not only concentrate 80 percent of the ownership of power generation in Chile, but also flood nearly 6,000 hectares and eventually construct 2,000 kilometers of transmission lines and mega-towers.[118] The massiveness of the public questioning of the Hydroaysén project is reflected in a recent survey that concluded that 74 percent of Chileans reject the dams and 50 percent are willing to pay higher electricity bills if necessary to stop their construction.[119] The scale of the Hydroaysén conflict shows that the factors that led to the institutional crisis that exploded in Valdivia in 2004 were not addressed by the 2009 reform.[120] Thus, political and institutional answers to the internal demand for environmental democratization are still pending in Chile.

This emergent demand for environmental democracy sharply contrasts with the external pressures that led to the establishment of the country's environmental law in 1994.[121] However, the diversity and contingent character of this apparently disconnected multitude of grassroots actors pose important obstacles for attempts of strategic articulation. Still, they represent the most vital manifestation of Chile's contemporary environmentalism. How they will eventually transcend their apparent disconnection or coordinate with traditional strands of Chilean environmentalism remain open questions.

Epilogue

In July 2013 a Valdivian court sentenced CELCO-ARAUCO to repair the damage caused to the Rio Cruces Sanctuary. The company did not appeal and accepted its responsibility. The restoration of the wetland is under design, and signs of recovery have been observed.

Acknowledgments

We thank Claudio Rivera and Arturo Vallejos, from the Universidad de Los Lagos, Chile, for their insightful comments.

Notes

1. Harvey, *A Brief Story of Neoliberalism.*

2. Aldunate, *El Factor Ecológico.*

3. Aldunate, *El Factor Ecológico.*

4. Rojas, "The Environmental Movement" and Clapp, "Waiting for the Forest Law."

5. Tecklin et al., "Making Environmental Law for the Market," and Silva, "Democracy, Market Economy."

6. See Geisse and Nelson, "De la Reacción a la Iniciativa," Geisse, "6to Encuentro Científico," Sabatini and Sepúlveda, *Conflictos Ambientales,* and Sepúlveda, "Cómo Fortalecer la Prevención de Conflictos?"

7. Geisse, "6to Encuentro Científico," Geisse and Nelson, "De la Reacción a la Iniciativa," Tecklin et al., "Making Environmental Law for the Market," Sepúlveda and Villarroel, "Swans, Conflicts and Resonance."

8. Carruthers, "Environmental Politics in Chile," Silva, "Democracy, Market Economy."

9. The first author of this chapter was a founder and active member of the environmental movement that emerged in Valdivia in response to this ecological disaster.

10. Sepúlveda and Villarroel, "Swans, Conflicts and Resonance."

11. Aldunate, *El Factor Ecológico,* Carruthers, "Environmental Politics in Chile."

12. Larraín, "Un Movimiento con Tres Corrientes," Aldunate, *El Factor Ecológico,* Fundación Terram, "Participación Ciudadana, Organizaciones No Ggubernametales Verdes y Agenda Política."

13. Carruthers, "Environmental Politics in Chile."

14. Muñoz, "Las ONG's Ambientales," Aldunate, *El Factor Ecológico.*

15. Aldunate, *El Factor Ecológico.*

16. Silva, "Democracy, Market Economy."

17. Aldunate, *El Factor Ecológico,* Carruthers, "Environmental Politics in Chile."

18. Aldunate, *El Factor Ecológico,* Camus and Hajek, *Historia Ambiental de Chile.*

19. Aldunate, *El Factor Ecológico.*

20. Stutzin, "Un Imperativo Ecológico."

21. Pinochet de la Barra, *El Pensamiento de Eduardo Frei.*

22. Camus and Hajek, *Historia Ambiental de Chile.*

23. Frei Montalva, "Mensaje Presidencial," Camus and Hajek, *Historia Ambiental de Chile.*

24. At least three spiritual initiatives that emerged in the 1960s became the matrix for a whole generation of environmental leaders a decade after. These groups were the Silo humanist community, the Arica spiritual Group, and the Lola Hoffman's spiritual community.

25. Some of its members included Sara Larraín, Malú Sierra, Ximena Abogabir, Andrés Wood, Vittorio Di Girólamo, Francisco Hunneus, Pía Figueroa, and Patricia Politzer.

26. Aldunate, *El Factor Ecológico.*

27. Aldunate, *El Factor Ecológico.*

28. Silva, "Democracy, Market Economy."

29. Clapp, "Waiting for the Forest Law," Estenssoro, *Medio Ambiente e Ideología.*

30. Clapp, "Waiting for the Forest Law," 24–25.

31. These NGOs include, for example, the Centro de Educación y Tecnología, CET (Center for Education and Technology), Canelo de Nos, and the Centro de Experimentación y Capacitación en Tecnología Apropiada, TECKNE (Centre for the Experimentation and Training on Appropriate Technology).

32. Larraín, "Un Movimiento con Tres Corrientes," Aldunate, *El Factor Ecológico,* Estenssoro, *Medio Ambiente e Ideología.*

33. Aldunate, *El Factor Ecológico,* Carruthers, "Environmental Politics in Chile."

34. Baquedano, "Los Ecologismos y la Política en Chile." See also, La Época, "Recién creado Instituto de Ecología Política."

35. Aldunate, *El Factor Ecológico.*

36. Rojas, "The Environmental Movement," Clapp, "Waiting for the Forest Law," Saavedra, "Nuestra Comunidad Científica."

37. Larraín, "Un Movimiento con Tres Corrientes," Aldunate, *El Factor Ecológico.*

38. Aldunate, *El Factor Ecológico.*

39. Aldunate, *El Factor Ecológico.*

40. Aldunate, *El Factor Ecológico.*

41. Biblioteca del Congreso Nacional, "Los Verdes."

42. Aldunate, *El Factor Ecológico.*

43. "Our call is not to political mentalities, but to the citizen, the neighbour, the brave . . . We are human beings that want to inhabit an inhabitable world and who are not willing to leave in the hands of politicians or the military-industrial complexes . . . We invite to confrontation through active non-violence against all who threatens life . . ." (Biblioteca del Congreso Nacional, "Los Verdes.")

44. Marambio, "Organizaciones No Gubernamentales Ambientalistas."

45. Biblioteca del Congreso Nacional, "Los Verdes."

46. CONICYT, *Principios para una política ambiental.*

47. CIPMA, "Propuesta para Iniciar," CIPMA, *Bases para una Ley del Medio Ambiente.*

48. Aldunate, *El Factor Ecológico.*

49. Carruthers, "Environmental Politics in Chile," Silva, "Democracy, Market Economy."

50. Carruthers, "Environmental Politics in Chile," 348.

51. Tecklin et al., "Making Environmental Law for the Market," Silva, "Democracy, Market Economy."

52. Carruthers, "Environmental Politics in Chile," 348.

53. Rojas et al., "Conflictos Ambientales en Chile."

54. Silva, "Democracy, Market Economy," 15.

55. Carruthers, "Environmental Politics in Chile," Silva, "Democracy, Market Economy."

56. Tecklin et al., "Making Environmental Law for the Market."

57. Chile's incorporation to the NAFTA failed in the U.S. Congress. However, the country signed a free trade agreement with Canada in 1996 and with Mexico in 1998. A bilateral agreement was also reached with the United States in 2003. See also Tecklin et al., "Making Environmental Law for the Market," 885.

58. Silva, "Democracy, Market Economy."

59. Tecklin et al., "Making Environmental Law for the Market," 887.

60. Tecklin et al., "Making Environmental Law for the Market," 887.

61. Tecklin et al, "Making Environmental Law for the Market," Sepúlveda and Villarroel, "Swans, Conflicts and Resonance," Geisse, "6to Encuentro Científico."

62. Tecklin et al., "Making Environmental Law for the Market," Silva, "Democracy, Market Economy."

63. Tecklin et al., "Making Environmental Law for the Market."

64. Sepúlveda, "¿Cómo Fortalecer la Prevención," Sepúlveda and Villarroel, "Swans, Conflicts and Resonance."

65. Sepúlveda and Villarroel, "Swans, Conflicts and Resonance."

66. Sepúlveda and Bettati, "El Desastre Ecológico del Santuario del Río Cruces," WWF, "Informe de Observaciones y Recomendaciones."

67. Sabatini and Sepúlveda, *Conflictos Ambientales.*

68. Sepúlveda, "¿Cómo Fortalecer la Prevención."

69. Sabatini and Sepúlveda, *Conflictos Ambientales.*

70. Reyes and Sepúlveda, forthcoming.

71. Reyes and Sepúlveda, forthcoming.

72. Camus and Hajek, *Historia ambiental de Chile.*

73. Camus and Hajek, *Historia ambiental de Chile,* Clapp, "Waiting for the Forest Law."

74. Sepúlveda, "¿Cómo Fortalecer la Prevención."

75. Sepúlveda and Du Monceau, "El Comportamiento del SEIA."

76. Rojas et al., "Conflictos Ambientales en Chile."

77. Sepúlveda and Villarroel, "Swans, Conflicts and Resonance."

78. Rojas et al., "Conflictos Ambientales en Chile," Sepúlveda, "Ciudad de Papel," Sepúlveda and Villarroel, "Desastre Ecológico de CELCO."

79. OEDC, "Environmental Performance Reviews—Chile."

80. Latta and Cid, "Testing the Limits."

81. Rojas et al., "Conflictos Ambientales en Chile."

82. Sepúlveda, "La Doctrina Frei."

83. Sepúlveda and Villarroel, "Swans, Conflicts and Resonance."

84. WWF, "Informe de Observaciones y Recomendaciones," UACH, "Estudio sobre Origen de Mortalidades."

85. Sepúlveda and Bettati, "El Desastre Ecológico del Santuario del Río Cruces," Pizarro, "Los Cinco Problemas de la Institucionalidad Ambiental," Larraín, "Desafíos Ambientales del Desarrollo Nacional," Sepúlveda and Villarroel, "Desastre Ecológico de CELCO."

86. WWF, "Informe de Observaciones y Recomendaciones."

87. UACH, "Estudio sobre Origen de Mortalidades," CONAF, "Plan Integral de Gestión Ambiental."

88. Lagos, "Variabilidad de Eventos de Mortalidad."

89. Sepúlveda and Mariángel, "La Legitimidad del Sistema de Evaluación de Impacto."

90. Sepúlveda and Mariángel, "La Legitimidad del Sistema de Evaluación de Impacto."

91. Acción por los Cisnes, "Fundamentos de la Agrupación Ciudadana."

92. Acción por los Cisnes, "Fundamentos de la Agrupación Ciudadana."

93. COREMA, "Resolución Exenta 017."

94. UACH, "Estudio sobre Origen de Mortalidades."

95. SISS, "Ordinario 505."

96. COREMA, "Resolución Exenta 377."

97. Alvarado, "Informe Pericial," Victoriano, "Informe Pericial."

98. Sepúlveda and Villarroel, "Swans, Conflicts and Resonance."

99. These business actors had only agreed to back minor changes aimed to facilitate the approval of projects by the SEIA, which were approved in 2002 during Ricardo Lagos's administration.

100. OECD, "Environmental Performance Reviews—Chile."

101. Bachelet, "Estoy Contigo."

102. Gobierno de Chile, "Mensaje Presidencial no. 352–356."

103. Sepúlveda and Villarroel, "Swans, Conflicts and Resonance."

104. Cámara de Diputados, "Oficio No 8080 del 6 de mayo de 2009."

105. Sepúlveda and Rojas, "Conflictos Ambientales y Reforma Ambiental en Chile."

106. Sepúlveda and Rojas, "Conflictos Ambientales y Reforma Ambiental en Chile."

107. Sepúlveda and Rojas, "Conflictos Ambientales y Reforma Ambiental en Chile."

108. Sepúlveda and Villarroel, "Swans, Conflicts and Resonance."

109. La Tercera, "Chile Exige que el Gobierno Cumpla," Latta and Cid, "Testing the Limits."

110. Sepulveda and Rojas 2010, Latta and Cid, "Testing the Limits."

111. Sepúlveda and Villarroel, "Swans, Conflicts and Resonance."

112. Aldunate, *El Factor Ecológico.*

113. The Observatorio Ciudadano was originally founded in Temuco in 2004 as Observatorio de Pueblos Indígenas (Indigenous People's Watchdog), changing to its current name in 2008.

114. Mori, "Anhelos, Perspectivas y Visión de Futuro," Feedback, "Percepciones y Actitudes hacia el Tema Medio Ambiental," TNC, "Encuesta sobre Conservación," La Tercera, "75% Rechaza Hidroaysén."

115. Mori, "Anhelos, Perspectivas y Visión de Futuro," Feedback, "Percepciones y Actitudes hacia el Tema Medio Ambiental."

116. TNC, "Encuesta sobre Conservación."

117. More than 80,000 persons marched only in Santiago against the approval of the Hidroaysén project on May 28, 2011 (Radio Cooperativa, "Minuto a Minuto: Revise la manifestación contra HidroAysén"). This march was considered the most massive one since the return of democracy in 1990 and before the student movement that exploded that same year (Mendoza, "El NO a Hidroaysén").

118. Latta and Cid, "Testing the Limits."

119. La Tercera, "75% Rechaza Hidroaysén."

120. Sepúlveda and Villarroel, "Swans, Conflicts and Resonance."

121. Sepúlveda and Villarroel, "Swans, Conflicts and Resonance."

Bibliography

Acción por los Cisnes. "Fundamentos de la agrupación ciudadana Acción por los Cisnes en relación al desastre ecológico en el Santuario de la Naturaleza del Río Cruces." Valdivia, Chile: Acción por los Cisnes, 2005.

Aldunate, C. *El factor ecológico: las mil caras del pensamiento verde.* Santiago: LOM, 2001.

Alvarado, J. "Informe Pericial. Causa Rol # 746–2005 Caratulada Estado-Fisco con Celulosa Arauco y Constitución S.A." Valdivia, Chile: Primer Juzgado Civil de Valdivia, 2011.

Bachelet, M. "Estoy contigo: Programa de Gobierno 2006–2010." Santiago: Concertación de Partidos por la Democracia, 2005.

Baquedano, M. "Los ecologismos y la política en Chile." *Diario La Época,* November 14, 1991.

Biblioteca del Congreso Nacional. "Los Verdes." In *Historia Polírica Legislativa del Congreso Nacional de Chile.* http://historiapolitica.bcn.cl/partidos_politicos/wiki/Los_Verdes (Accessed June 11, 2012).

Cámara de Diputados. "Oficio No 8080 del 6 de mayo de 2009. Informe del Primer Trámite Constitucional del proyecto de ley que crea el Ministerio de Medio Ambiente, la Superintendencia de Medio Ambiente y el Servicio de Evaluación Ambiental." Enviado al Presidente del Senado por la Cámara de Diputados. Valparaíso: Congreso de Chile, 2009.

Camus, P., and E. Hajek. *Historia ambiental de Chile.* Santiago: Ediciones Pontificia Universidad Católica de Chile, 1998.

Carruthers, D. "Environmental Politics in Chile: Legacies of Dictatorship and Democracy." *Third World Quarterly* 22, no. 3 (2001): 343–58.

CIPMA (Centro de Investigación y Planificación del Medio Ambiente). *Bases para una ley del medio ambiente en Chile.* Santiago: CIPMA, 1993.

CIPMA (Centro de Investigación y Planificación del Medio Ambiente). "Propuesta para iniciar un proceso participativo hacia una política nacional de medio ambiente." *Ambiente y Desarrollo* 7, no. 1 (1991): 41–101.

Clapp, R. "Waiting for the Forest Law: Resource-led Development and Environmental Politics in Chile." *Latin American Research Review* 33, no. 2 (1998): 3–36.

CONAF (Corporación Nacional Forestal). "Plan integral de gestión ambiental del humedal de Río Cruces: Resumen ejecutivo, Agosto de 2006." Valdivia, Chile: Gobierno Regional Décima Región, 2006.

CONICYT. *Principios para una política ambiental.* Comité de Ciencias Ambientales. Santiago: CONICYT, 1988.

COREMA (Comisión Regional del Medio Ambiente). "Resolución Exenta 017, del 18 de enero de 2005." Puerto Montt, Chile: Gobierno Regional Décima Región, 2005a.

COREMA (Comisión Regional del Medio Ambiente). "Resolución Exenta 377, 6 de junio de 2005." Puerto Montt, Chile: Gobierno Regional Décima Región, 2005b.

Estenssoro, F. *Medio Ambiente e Ideología. La Discusión pública en Chile, 1992–2002. Antecedentes para una historia de las ideas políticas a inicios del siglo XXI.* Santiago: Ariadna, USACH, 2009.

FEEDBACK. "Percepciones y actitudes hacia el tema medio ambiental en Puerto Montt." Santiago: FEEDBACK/CEAM, Universidad Austral de Chile, 2005.

Frei Montalva, E. "Mensaje Presidencial." Santiago: Presidencia de Chile, 1966.

Fundación Terram. "Participación ciudadana, organizaciones no gubernametales verdes y agenda política." Documento de Trabajo, 1999.

Geisse, Guillermo. "6to Encuentro Científico sobre el Medio Ambiente: Demanda nacional por el medio ambiente y la calidad de vida." *Ambiente y Desarrollo* 13, no. 4 (1997): 4–6.

Geisse, Guillermo, and Michael Nelson. "De la reacción a la iniciativa: política ambiental chilena." *Ambiente y Desarrollo* 11, no. 2 (1995): 7–11.

Gobierno de Chile. "Mensaje presidencial no. 352–356, 5 de junio de 2008." Santiago: Presidencia de la República, 2008.

Harvey, D. *A Brief Story of Neoliberalism.* Oxford: Oxford University Press, 2007.

La Época. "Recién creado Instituto de Ecología Política intentará formar un movimiento verde chileno." August 17, 1987.

Lagos, N. "Variabilidad de eventos de mortalidad y reproductivos del cisne de cuello negro en el Santuario de la Naturaleza Carlos Anwandter (Río Cruces, Valdivia) durante el periodo 2003 a 2005." Santiago: Consejo de Defensa del Estado, 2008.

Larraín, S. "Desafíos ambientales del desarrollo nacional: Evaluación desempeño 1997–2006 y propuesta institucional." Serie e Foco 92. Santiago: Expansiva, 2006.

Larraín, S. "Un movimiento con tres corrientes." *La Nación,* December 13, 1990.

La Tercera. "Chile exige que el gobierno cumpla. Organizaciones ciudadanas, sociales y sindicales demanda cumplimiento de compromisos ambientales a gobierna de la Presidenta Bachelet." Inserto. November 10, 2009.

La Tercera. "75% rechaza Hidroaysén." May 15, 2011.

Latta, A., and B. Cid. "Testing the Limits: Neoliberal Ecologies from Pinochet to Bachelet." *Latin American Perspectives* 39, no. 4 (2012): 163–80.

Marambio Thibaut, M.C. "Organizaciones no gubernamentales ambientalistas y partidos políticos verdes, frente a la contaminación atmosférica. análisis comparativo: Chile y México, a partir de 1992." Tesis para obtener el grado de Maestra en sociología. Mexico, DF: Universidad Iberoamericana, 2004.

Mendoza, M. "El NO a Hidroaysén: ¡Es la gente estúpido!." *El Mostrador,* May 21, 2011. http://www.elmostrador.cl/opinion/2011/05/21/el-no-a-hidroaysen-¡es-la-gente-estupido/ (Accessed May 21, 2011).

MORI (Market Opinion Research International). "Anhelos, perspectivas y visión de futuro de los habitantes de la X Región. Resumen Ejecutivo Estudio enero 1999." Santiago: MORI, Chile, 1999.

Muñoz, A. M. "Las ONG's ambientales, actores fundamentales de la gestion ambiental." In *Nuevo gobierno: desafíos de la reconciliación 1999–2000,* edited by FLACSO. Santiago: FLACSO-Chile, 2000.

OECD (Organization for Economic Co-operation and Development). "Environmental Performance Reviews—Chile." Paris: OECD/CEPAL, 2005.

Pinochet de la Barra, O. *El Pensamiento de Eduardo Frei.* Santiago: Editorial Aconcagua, 1983.

Pizarro, R. "Los cinco problemas de la institucionalidad ambiental de Chile." Serie en Foco 89. Santiago: Expansiva, 2006.

Radio Cooperativa. "Minuto a Minuto: Revise la manifestación contra HidroAysén en la Alameda." May 28, 2011. http://www.cooperativa.cl/prontus_nots/site/artic/20110528/pags/20110528150807.html (Accessed June 25, 2011).

Reyes, R., and C. Sepúlveda. "Gobernanza de los bosques en Chile: entre el mercado internacional y las demandas ciudadanas." In upcoming book, edited by C. Donoso, title forthcoming.

Rojas, A. "The Environmental Movement and the Environmentally Concerned Scientific Community in Chile." *European Review of Latin American and Caribbean Studies* 56 (1994): 93–118.

Rojas, A., F. Sabatini, and C. Sepúlveda. "Conflictos Ambientales en Chile: aprendizajes y desafíos." *Ambiente y Desarrollo* 19, no. 2 (2003): 22–30.

Saavedra, I. "Nuestra comunidad científica y sus responsabilidades locales y globales." *Ambiente y Desarrollo* 1, no. 1 (1984): 15–18.

Sabatini, F., and C. Sepúlveda, eds. *Conflictos ambientales: Entre la globalización y la sociedad civil.* Santiago: CIPMA, 1997.

Sepúlveda, C. "Ciudad de Papel." Valdivia, Chile: Jirafa Films, 2007.

Sepúlveda, C. "¿Cómo fortalecer la prevención de conflictos en el SEIA?: recomendaciones a partir de la percepción ciudadana." *Ambiente y Desarrollo* 15, no. 4 (1999): 11–23.

Sepúlveda, C. "La doctrina Frei de los impactos ambientales consumados: el karma que a Bachelet le toca superar." *Valdivia Noticias,* June 25, 2006. http://www.valdivianoticias.cl (Accessed June 25, 2006).

Sepúlveda, C., and B. Betatti. "El desastre ecológico del Santuario del Río Cruces: Trizadura institucional y retroceso democrático." *Ambiente y Desarrollo* 20–21, no. 3–1 (2005): 62–68.

Sepúlveda, C., and M.I. Du Monceau. "El comportamiento del SEIA ante una comunidad que se toma la participación en serio: El Proyecto Costanera Norte." *Ambiente y Desarrollo* 14, no. 3 (1998): 23–33.

Sepúlveda, C., and P. Mariángel. "La legitimidad del Sistema de Evaluación de Impacto Ambiental puesta en juego: el caso de la Planta de Celulosa Valdivia." *Ambiente y Desarrollo* 14, no. 2 (1998): 6–17.

Sepúlveda, C., and A. Rojas. "Conflictos ambientales y reforma ambiental en Chile: una oportunidad desapro- vechada de aprendizaje institucional sobre participación ciudadana." *Ambiente y Desarollo* 24, no. 2 (2010): 15–23.

Sepúlveda, C., and P. Villarroel. "Desastre ecológico de CELCO en el Santuario Río Cruces: trizadura institucional y retroceso democrático." In *Conflictos por el agua en Chile: Entre los derechos humanos y las reglas del Mercado,* edited by Sara Larraín and Pamela Poo, 318–38. Santiago: Programa Chile Sustentable, 2010.

Sepúlveda, C., and P. Villarroel. "Swans, Conflicts and Resonance: The Role of Local Movements in the Reform of Chilean Environmental Institutions," *Latin American Perspectives* 39, no. 4 (2012): 181–200.

Silva, E. "Democracy, Market Economy and Environment Policy in Chile." *Journal of Interamerican Studies and World Affairs* 38, no. 4 (1996): 1–33.

SISS (Superintendencia de Servicios Sanitarios). "Ordinario 505, dirigido a la Gerencia de la Planta Valdivia de CELCO-ARAUCO, 24 de marzo de 2005." Puerto Montt, Chile: Superintendencia de Servicios Sanitarios, Décima Región, 2005.

Stutzin, G. "Un imperativo ecológico: reconocer los derechos a la naturaleza." *Ambiente y Desarrollo* 1, no. 1 (1984): 97–114.

Tecklin, D., C. Bauer, and M. Prieto. "Making Environmental Law for the Market: The Emergence, Character, and Implications of Chile's Environmental Regime." *Environmental Politics* 20 (2011): 879–98.

TNC (The Nature Conservancy). "Encuesta sobre conservación de biodiversidad y áreas protegidas en Chile." Santiago: TNC, 2009.

UACH (Universidad Austral de Chile). "Estudio sobre origen de mortalidades y disminución poblacional de aves acuáticas en el Santuario de la Naturaleza Carlos Anwandter, en la Provincia de Valdivia, informe final. Convenio Complementario Especifico 1210–1203/2004–12–14." Valdivia, Chile: UACH, 2005.

Victoriano, P. "Informe Pericial. Causa Rol # 746–2005 Caratulada Estado-Fisco con Celulosa Arauco y Constitución S.A." Valdivia, Chile: Primer Juzgado Civil de Valdivia, 2011.

WWF (World Wildlife Fund). "Informe de observaciones y recomendaciones, Misión Internacional de Evaluación de WWF ante la controversia del Santuario de la Naturaleza y sitio Ramsar Carlos Anwandter y la planta de celulosa Valdivia de CELCO-ARAUCO." Valdivia, Chile: WWF, 2005.

Chapter 3

Dam-Nation: The Struggle over the Future of the Brazilian Amazon

Benito Cao

> I think that today the war is about to start once more and the Indians will be forced to kill the white men again so they leave our lands alone.
>
> —Chief Raoni (2010)

Brazil is a country of critical importance to the global environment in the 21st century. The world's fifth largest territory (8.5 square kilometers), with the fifth largest population (183 million), Brazil is one of the five major emerging economies, the B in the BRICS, alongside Russia, India, China, and South Africa. Brazil is also one of the top 10 emitters of greenhouse gases, most of which come from the Brazilian Amazon, a region that comprises 60 percent of the total Amazon. The Brazilian Amazon contains roughly 40 percent of the world's remaining tropical rainforest, the preservation of which has become a global political priority because of its significance for biodiversity and the mitigation of climate change. The ecological significance of the Brazilian Amazon is also reflected in the attention it has received from environmental activists and nongovernmental organizations (NGOs), especially since the 1980s. Conversely, the prominent position the Amazon holds in the national and global imagination makes the struggle over the ecological future of this region symbolic of the strength of environmental activism in and beyond Brazil.

This chapter provides an overview of the history of Brazilian environmentalism, with a particular focus on its role in protecting the Amazon. Focusing on this region will enable us to explore the links between the local, national, and global dimensions of environmental activism in Brazil. At the same time, it will expose the civilizing discourse underpinning the integration of the Amazon in the Brazilian imaginary.

The current state of the Amazon reflects the tension between environment and development, between ecology and economy, that has shaped much of the global political agenda, especially since the concept of sustainable development was officially embraced at the Earth Summit of 1992 held in Rio de Janeiro, Brazil. This tension is exacerbated by the long-held Brazilian fear regarding the so-called internationalization of the Amazon—the notion that foreign powers are always conspiring to occupy the Amazon. The global concern with climate change adds to these fears, especially when the Amazon is referred to as being too important for the planet to be left exclusively in the hands of Brazilians. These dynamics have deep-seated roots in the history of Brazil and are all at play in the case study explored in the second part of this chapter, the struggle over the Belo Monte dam.

Exploring the Past: Forests, Activists, and Networks

> Conquering the land, dominating the waters, subjugating the forest have been our tasks. And, in that struggle, which has been going on for centuries, we are achieving triumph after triumph.
>
> —Getúlio Vargas (1941)

The Amazon occupies a special place in the global imagination but is also central to the national history of Brazil. The Brazilian Amazon occupies over half of the national territory, comprising the states of Acre, Amapá, Amazonas, Pará, Rondônia, Roraima, Tocantins, Mato Grosso, and most of Maranhão. This area is known in Brazil as the "Legal Amazon." The Amazon holds the world's largest river system, the world's largest rainforest, a fifth of the world's freshwater, and produces roughly one-third of the world's oxygen, over an area larger than Europe.[1] The Amazon is also one of the most critical reservoirs of biodiversity on the planet, home to an estimated 25 percent of animal and plant species found on Earth, many of them unique to the Brazilian Amazon.[2] Contrary to what most people think, one-third of the Brazilian Amazon is covered by grassland and savannahs, but it is the monumental size of its rainforest that inspired the notion of the Amazon as the "lungs of the earth," and situated the region at the forefront of the environmental movement, both globally and in Brazil.

This preoccupation with the forest has often come at the expense of ignoring the indigenous peoples that have populated this region for centuries.

The invisibility of the indigenous peoples of Brazil, also known as Brazilian Indians, is connected to the notion that the Amazon was a pristine forest, undisturbed by human activity, until the arrival of the Europeans. Increasing evidence indicates that this "pristine myth"[3] does not reflect the historical reality of the Amazon. Instead, recent studies conceptualize the Amazon as a "cultural parkland,"[4] a "constructed landscape,"[5] in which patterns of biodiversity have been largely shaped by its indigenous peoples.[6] The myth of a pristine forest goes hand in hand with depictions of Brazilian Indians as primitive peoples.[7] These narratives effectively erase past indigenous agency and deny future indigenous agency when it comes to shaping the Amazon. The pristine myth privileges the agency of natural scientists, whilst the "primitive myth" privileges the agency of social scientists. As a result, the tension between conservation and development is often played out sidelining the agency of Brazilian Indians, both in the efforts to protect the Amazon forest and in the efforts to develop the Amazon region. These dynamics are further complicated by the special place the Amazon occupies in the formation and formulation of Brazil.

The history of the Brazilian Amazon is a history of the gradual occupation and appropriation of this region first by the Portuguese colonists and then by the Brazilian state.[8] It is also a history of exploration and exploitation, often aided by the use of slave labor, both indigenous and of African origin.[9] It is worth noting that in Portuguese, the same word, namely *exploração,* means both exploration and exploitation. This linguistic curiosity betrays the nature of the Portuguese colonization of Brazil and even more so of their inroads into the Amazon. Similarly, and tellingly, the centrality of the motto "Order and Progress" in the Brazilian flag reflects the modernist ideals that underpinned the country's transformation into a republic in the late 19th century, and more generally, it illustrates the civilizing discourse that dominates Brazilian history, particularly regarding the integration of the Amazon.

The civilizing and modernizing impulse with regard to the Amazon intensified in the first half of the 20th century, beginning with the pioneer expedition led by Marechal Cândido Rondon in 1907,[10] continuing with the March to the West launched by Getúlio Vargas in 1938,[11] and culminating with the transferring of the country's capital from Rio de Janeiro to Brasilia in 1960. These initiatives were all efforts to bring the Amazon and its peoples under the effective, that is, not just legal control of the Brazilian state.[12] They were also part of a nation-building project based on the views of the Amazon as a repository of wealth to be exploited for the benefit of the Brazilian people. The Amazon was to be integrated into the national economy and its resources used to power the economic growth of Brazil.[13]

The dynamics established by civilian governments in the first half of the 20th century intensified under the military dictatorship that ruled Brazil between 1964 and 1986, producing the so-called economic miracle of the 1970s and marking the beginning of "the onslaught on the Amazon."[14]

The impact of the modernizing impulse on the natural environment led to the first organized expressions of concern over the state of the environment in Brazil. This included the First Brazilian Conference on Nature Protection, held in Rio de Janeiro, in 1934[15] and the establishment of the first environmental association, in 1954, also in Rio de Janeiro: the Brazilian Foundation for the Conservation of Nature (FBCN). The FBCN was a strictly conservationist group formed by an elite group of naturalists, many of them civil servants with ties to the Brazilian political elite.[16] The association operated as an interest group, seeking to influence policy decisions directly, without building alliances with political parties or social movements. This gave the group "the appearance of a parallel government organization until the 1970s."[17] The FBCN contributed to the establishment of protected areas, such as national parks, and to the elaboration of the Forest Code of 1965.[18] However, their work had little impact on the breakneck pace at which the process of industrialization was proceeding in Brazil.

The approach to the Amazon during the military regime was framed by two telling slogans of the time: *integrar para não entregar* (integrate it, don't abdicate it) and *uma terra sem homens para homens sem terra* (a land without men for men without land). The first slogan was a reflection of the military's deep fear of losing control of the Amazon. This historical fear was being exacerbated at the time by references to Brazil's "relative sovereignty" over the Amazon, made during the Stockholm Conference of 1972.[19] The second slogan encapsulates the state's response to the issue of poverty in the northeast. Rather than pursuing land reform, landless peasants were to be settled on "land without men"—furthering the colonization of the Amazon.[20] These slogans betray the invisibility of the Brazilian Indians, and showcase the vision of the Amazon as a space crucial to the geopolitical and economic future of Brazil.

The slogans were operationalized through a series of development plans, including Operation Amazonia (1964) and the National Integration Plan (1970), under the coordination of the Superintendency for the Development of Amazonia (SUDAM). The most ambitious project was the Transamazonian Highway (BR-230), a 5,300 kilometer road that would link the Brazilian northeast with the capital Brasilia and pierce the forest all the way to the border with Peru.[21] The highway, inaugurated in 1972, has been one of the main drivers of the deforestation of the Amazon, along with the network of roads built in the 1960s (BR-010, BR-153, BR-364) that carved the infamous "arc of deforestation"—the crescent-shaped area along the eastern and southern edges of the Amazon forest where 80 percent of the deforestation is concentrated.[22]

The environmental impact of the policies driving the economic miracle of the 1970s increased the concern for the environment, which now extended to other sectors of the middle class, including artists, lawyers, journalists, and businessmen. This led to the transformation of environmentalism from a loose network of individuals (i.e., concerned scientists operating from inside official institutions) into a social movement that spread rapidly across the nation, replacing the narrow corridors of power with the streets as the main stage for environmental activism in Brazil.

The first and most prominent groups in the early years of the movement were the Southern Association for the Protection of the Natural Environment (Agapan), founded in 1971, the Art and Ecological Thinking Movement (Mape), formed in 1973, and the São Paulo Association for Natural Protection (APPN), created in 1976. These groups reacted directly against the ideology of rapid and predatory growth that had taken root, epitomized by the invitation proffered by the government to first-world pollution-producing industries to move their operations to Brazil.[23] The three groups shared a broad concern for the environment, but differed in their specific identities, actions, and frames regarding the environment and environmental activism. Agapan focused primarily on the impact of public policy on the flora and fauna, whilst Mape and APPN were mostly preoccupied with the impact of public policy on the Brazilian people. Their repertoire of strategies was broad, including: campaigns, petitions, open letters, public lectures, literary events, exhibitions, and entertainments.[24] Crucially, these groups began to connect environmental issues with political debates, effectively transforming the environment into a "political question."[25]

On the whole, the environmental movement of the 1970s tended to balance conservationism and humanism, thus signaling a shift in the definition of environmental problems away from the conservationist frame that had dominated until then, and toward what came to be known as socio-environmentalism.[26] The environment was redefined—the image of an untouched natural world was replaced by one that emphasized the relationship between natural resources and social groups—and environmental problems came to be associated with the "development model" pursued by the military regime. This shift from an ecological toward an ecosocial approach brought together environmental activism and the movement for re-democratization that was gathering pace in the late 1980s.[27]

The event that projected the environmental movement onto the national stage was the Campaign in Defense of the Amazon, launched in 1978. The campaign opposed the federal government's plan to draw up contracts with international companies for the extraction of timber from the Amazon forest.[28] The campaign became the Movement in Defense of the Amazon, with local committees in 18 states and the federal district. The participants exchanged documents nationally, held numerous demonstrations, carried out letter-writing campaigns, and organized conferences to educate themselves and local populations. They also organized the First National Congress in Defense of the Amazon, held in Brasilia in 1980. The most memorable moment of the campaign came on January 15, 1979, when a public act in São Paulo attracted 1,500 people and spawned the "Open letter to the Brazilian nation." The letter reflected popular concerns with the exploitation and sovereignty of the Amazon.[29]

The ecosocial approach extended the appeal of environmentalism beyond its traditional middle-class constituency, attracting the urban popular classes and the rural poor, particularly in the Amazon region. In the urban areas, the alliance between environmentalists and social activists translated into campaigns dealing with issues such as waste disposal, health care, and industrial pollution. The residents of the urban periphery began to participate in politics using environmental arguments—a phenomenon which has been termed "indirect environmentalism."[30] In the rural areas, the main concern was deforestation, especially in the Amazon region, where the formidable figure of Chico Mendes came to the fore. His passionate defense of the rubber tappers and of the Amazon forest from where they drew their livelihoods exemplified a "Trees and People" approach[31] that repositioned the rubber tappers as defenders of the Amazon forest, embedding their concerns and activism into the environmental movement that was gathering strength in Brazil.

The adoption of the ecosocial approach was complemented by the addition of an ecopolitical dimension in the second half of the 1980s.[32] The unprecedented opportunity to influence the content of the new constitution added renewed vigor to Brazilian environmentalism in the years leading to the Constitutional Assembly. The decision to enter directly the political arena led to the creation of the Green Party, in 1986, and the establishment of the Green List—a strategy for supporting candidates who pledged their commitment to a list of green proposals. None of the Green Party candidates was elected, but one from the Green List was: Fabio Feldmann, from São Paulo. Feldmann convened a meeting of environmentalists in Brasilia, out of which came the National Front for Ecological Action in the

Constituent Assembly. In total, nine senators and 82 federal deputies signed up to support the environmental cause in the Constitutional Assembly.[33]

The pressure from civil society groups and parliamentary alliances, facilitated by the presence of Feldmann inside the assembly, resulted in the inclusion of a chapter on the environment in the Constitution (Chapter VI). The chapter ensures the right to a balanced environment and imposes on the government and the people the duty to defend and preserve that balance. The government must also foster ecological education at all levels and promote public awareness of the need to preserve the environment. The Constitution declared several natural areas national treasures, assigning responsibility for their protection to the federal government—among these national treasures is the Amazon forest.

The Constitution also included a chapter (Chapter VIII) recognizing the cultural and territorial rights of indigenous peoples, establishing their right to live in traditional territories, and guaranteeing the "exclusive use" of the natural resources in their lands (art. 231, section 2). The chapter allows for the exploitation of hydro resources and mineral riches in those lands, but only "with the authorization of the National Congress" and "after hearing the communities involved" (art. 231, section 3). The chapter also grants indigenous people, their communities and organizations, the right to "sue to defend their rights and interests" (art. 232).

The combined effect of Chapter VI (Environment) and Chapter VIII (Indians) has had significant implications for environmental activism in the Amazon. The chapters provide environmentalists with a legal avenue of contestation against policies that damage the environment, especially those that impact on areas inhabited by indigenous peoples, most of which are in the Brazilian Amazon. The chapters have also made Brazilian Indians central to the policy process for the Amazon, at least formally. Their need to be heard along with their right to sue makes for a powerful legal combination at the disposal of the indigenous peoples of Brazil. This also positions them at the center of environmental activism when it comes to shaping the future of the Amazon.

The period of the Constitutional Assembly coincided with a series of trips abroad, sponsored by international NGOs, which took Chico Mendes to the United States and saw several indigenous leaders tour Europe and the United States raising awareness about the deforestation of the Amazon. During these trips, they met with government officials, political representatives, and executives of multilateral banks, such as the World Bank. Their lobbying efforts abroad translated into international pressure on the Brazilian authorities back at home—producing a "boomerang effect."[34]

The murder of Chico Mendes in 1988 by local ranchers created an "ecological martyr" and gave a human face to the deforestation of the Amazon.[35] His death, coupled with the massive fires in the Amazon forest, dramatically increased the pressure on the Brazilian authorities. In response, the government launched the Our Nature Project (1988), established the Brazilian Institute of the Environment and Natural Resources (IBAMA), in 1989, and put a bid to host the second United Nations Conference on the Environment and Development (UNCED) announced for 1992. The government also set up the Amazon Surveillance System (SIVAM), a network of radars, aircraft, satellites, and other sensors electronically tied to a centralized command center in Brasilia, to monitor the Amazon. SIVAM was to serve the dual purpose of assisting with the effort to curb deforestation as well as securing Brazilian sovereign control of the Amazon.[36]

The announcement of Rio de Janeiro as the host city for the conference created a frenzy of political activism across Brazil, leading to the formation of a broad network of environmental associations and social movements in 1990: the Forum of Brazilian NGOs and Social Movements for the Environment and Development. The Forum was a national network of over one thousand civil society organizations comprised mainly of two groups of similar size: environmental associations and groups without previous environmental activism, including popular movements, trade unions, professional associations, religious organizations, neighborhood groups, indigenous groups, feminist groups, student groups, and cultural groups.[37]

The Forum consolidated the network as the dominant operational structure of environmental activism in Brazil and produced a convergence of the two primary frames that had shaped Brazilian environmentalism: conservationism, dominant until the 1970s, and socio-environmentalism, dominant during the 1980s. The convergence came about through "the dilution of both frames": conservationism was redefined by replacing the idea of ecosystem with that of biodiversity, which came to include social groups living in forest areas; and socio-environmentalism redirected its focus toward the living conditions of communities interacting with the natural environment.[38] The convergence sidelined urban environmental concerns, such as waste disposal, pollution, and sanitation,[39] and produced a new frame, neoconservationism.[40]

The emphasis given to technology and scientific knowledge in Agenda 21, the main outcome document of the Earth Summit, and the establishment of funding programs for the conservation of large biomes strengthened this new frame, and has favored environmental organizations with strong research profiles and expert staff specialized in specific ecosystems, such as SOS Mata Atlântica, created in 1986, focused on the Atlantic forest, and the Socio-Environmental Institute (ISA), created in 1994, focused on the Amazon forest.[41] The possibility of managing environmental reserves and transnational grants has stimulated the transformation of these and similar organizations into professional bureaucracies, creating the conditions for their cooperation with government and corporations.[42] This has increased their ability to shape environmental policy, but it has also reduced their ability to question existing power structures and environmental injustices.

The professionalization and institutionalization of environmental activism has also been stimulated by the funding and collaboration opportunities presented by the increasing presence of international NGOs in Brazil, such as Greenpeace, Friends of the Earth (FoE), the World Wide Fund for Nature (WWF), and Conservation International (CI). Their influence has contributed to add a global dimension to the environmental networks in Brazil. The resulting glocalization has strengthened the movement, but not without some troubling side effects, especially when it comes to the Amazon.

The campaigns against deforestation organized by these NGOs, such as the 1990s' "Mahogany is Murder" campaign by WWF/FoE and the current "Slaughtering the Amazon" campaign by Greenpeace, deploy a market-oriented approach that promotes technical solutions (e.g., certification schemes, monitoring, and tracking) that can guarantee the ethical production and consumption of timber, beef, and leather from the Amazon.[43] This privileges the agency of the global consumer and commodifies the Amazon and its people, reducing the scope for alternative modes of living and development that are not driven by processes of industrialization and resource extraction and/or production for global markets.[44]

The government response to the increasing salience of the environment has also brought ambivalent outcomes regarding the ability of Brazilian environmentalism to shape the future of the Amazon, and with it the future of Brazil. The creation of the Ministry of the Environment, in 1992, constitutes the symbolic culmination of a long and arduous road that has positioned the environment, at least formally, at the heart of the Brazilian government. However, the transformation of environmental concerns and values into specific policies has reinforced the technical and expert-driven approach to the environment, contributing—somewhat paradoxically—to the depoliticization of Brazilian environmentalism. Moreover, the internal dynamics of the executive, especially in the past decade, have betrayed the extent to which other traditional ministries, particularly those concerned with matters of national security and economic development, still dominate policy outcomes, mainly when it comes to formulating policy concerning the Brazilian Amazon.

These dynamics are all currently at play in the most important struggle for the future of the Amazon, the Belo Monte dam. The outcome of this struggle will have major repercussions for the future of the Brazilian Amazon and of Brazilian environmentalism in the 21st century.

Back to the Future: Dams, Indians, and Avatars

> I am from your future to tell you what you are going to be like and you are not going to like it.
>
> —James Cameron (2010)

The Belo Monte dam is the flagship of the Growth Acceleration Plan (PAC), the master plan for the Brazilian economy launched by President Lula in 2007, and championed by the current president, Dilma Rousseff. The dam is the centerpiece of a hydroelectric complex proposed for the Xingu River, the longest of the many tributaries of the Amazon River. With an installed capacity of 11,233 megawatts, Belo Monte would be the world's third largest dam, behind the Three Gorges dam in China and the Itaipu dam on the border between Brazil and Paraguay. The PAC includes other large-scale capital-intense infrastructure projects, such as the paving of the Transamazonian Highway (BR-230) and the Cuiabá-Santarem Highway (BR-163), signaling the continuation of the developmentalist philosophy embraced by Brazilian governments throughout the 20th century.

The origins of the Belo Monte dam date back to the 1970s. The idea of building a hydroelectric complex in the Xingu River was first proposed by Brazil's state-owned electric utility company Electrobrás and its regional subsidiary Electronorte in 1975.[45] The proposal contemplated the construction of six dams, all of which were given indigenous names. The first phase contemplated the construction of two dams: Kararâo (later renamed Belo Monte) and Babaquara. The combined area to be flooded was estimated at 14,500 square kilometers and the complex would have five generating plants along the Xingu, with a total capacity of 14,700 megawatts. Based on that plan, an entire indigenous reserve, the Paquiçamba, would be flooded and some 6,000 people would have to be relocated. These two dams would be followed by four additional dams, flooding a total area of 22,000 square kilometers in the Xingu basin.[46]

In 1987, the proposal became part of the "2010 Plan," a massive hydroelectric plan that listed 297 dams across Brazil, 79 of them in the Amazon.[47] As soon as the plan became public, the dam complex contemplated for the Xingu met with strong opposition, particularly from the indigenous peoples of the region, and from national and international NGOs. Together they organized a media campaign, raising awareness about the environmental and social impact of the project. In 1988, two Kayapó leaders flew to Washington and directly lobbied with the main funding source of the project, the World Bank.[48] The same year, another Kayapó leader, Raoni Metuktire (Chief Raoni) toured Europe with popular singer Sting, calling for the protection of the rainforest, the demarcation of Kayapó lands,[49] and gathering support for the protest against the dam complex that was being organized for 1989 in Altamira.[50]

In 1989, the indigenous peoples of the Xingu, led by the Kayapó, organized the First Meeting of the Indigenous Nations of the Xingu and staged a massive rally in Altamira, attended by 3,000 demonstrators, including indigenous peoples, rural workers, national and international politicians, and environmental and human rights activists, all demanding the Xingu remain free from dams.[51] The event attracted a lot of media attention and produced one of the most powerful images in the history of Brazilian environmentalism, when a Kayapó woman, EJ Tuíra, placed her machete against the face of the Electronorte representative, José Antônio Muniz Lopes[52]—CEO of Electrobrás from 2008 to 2011. The photo made the front page of the *New York Times,* contributing to raise the profile of the struggle and sparking widespread condemnation of the project.

The alliance between environmentalists and indigenous peoples came at a point of heightened visibility for both these movements, and at a crucial moment for the World Bank, under pressure for its poor environmental record. Shortly after the Altamira protest, the World Bank announced that it was denying the loan to Electronorte, leaving the government with little option but to shelve the plan in 1990.

The success of the anti-dam campaign rested largely on the political skills of the Kayapó. Over the years, they have repeatedly proven to be "effective political players"[53] and masters at using new media.[54] Kayapó leaders, like Megaron Txucarramãe, have even been able to gain appointments to strategic positions within official institutions, such as the National Indian Foundation (FUNAI). The Kayapó emerged from this battle as a formidable political force and were embraced by the international environmental movement as environmental warriors, symbols of the resistance against the growth-at-all-cost model epitomized by the Karârao dam. Importantly though, the success in the first battle over the dam complex rested largely on their ability to make themselves the center of a wide and ethnically diverse network of alliances with indigenous and nonindigenous peoples from the region. This has enabled the emergence of a regional indigenous identity[55] that will be crucial for the success of the current battle over the Belo Monte dam.

The project to build a dam complex in the Xingu River was resurrected in 2001, following the revival of economic growth in Brazil in the 1990s. Since then, there has been a renewed interest in energy generation projects, intensified after energy shortages led to rationing for nine months between June 2001 and March 2002.[56] Not surprisingly, the government turned to hydroelectric power as the energy source that would power the country's economic growth into the 21st century. Hydroelectric energy accounts for more than 90 percent of the electricity produced in Brazil,[57] yet this represents only around 30 percent of the hydropower capacity of Brazil.[58] In other words, there is an enormous amount of hydroelectric potential still untapped, and half of that potential is in the Legal Amazon. The appeal of hydroelectricity is bolstered by its reputation as a form of "clean energy," essential to mitigate climate change in the 21st century.

Electronorte began to work on new designs for what would become the Belo Monte dam complex in 1998. In an apparent effort to draw attention away from previous indigenous resistance to the project, the company changed the names of the dams contemplated in the new plan: Karârao became Belo Monte, and Babaquara became Altamira. The other four dams are absent from public discussion, but the continued activity of the company's engineers in the locations in question indicates their plans have not been abandoned.[59] The new design for the Belo Monte dam has reduced the area to be flooded to 440 square kilometers (down from 1,225 square kilometers). The reduction has been achieved by diverting the flow of the river through a series of channels to the generation plant, thus eliminating the need to flood the whole Paquiçamba reserve.

The revised plan was released in 2002, accompanied by a propaganda campaign portraying hydropower as a clean and sustainable way to fuel economic growth and promote human development in Brazil. In addition, the government announced a raft of mitigation and compensation mechanisms, and presented the Belo Monte dam as a unique opportunity to develop and create employment in the region, one of the poorest areas of Brazil.[60]

The critics argue that the dam will affect not only those living in the area that will be flooded, but also those living along the large stretch of the river known as the Big Bend (*Volta Grande*). This area would be dried out as the water is diverted to the powerhouse, thus at least doubling the number of people affected by the dam—displacing between 20,000 and 40,000 people.[61] The change in the river level would have considerable ecological impact, as it constitutes an important ecological regulator in the region. The social impact and environmental damage in the region will be compounded by the arrival of thousands of people in search of work. The government argues that the construction of the dam will generate 18,000 direct and 23,000 indirect jobs, but estimates the arrival in the region of 100,000 people. Experts argue that this large influx of people, many without jobs, will incense conflict over lands and natural resources and will fuel the deforestation of the Xingu region,[62] as has happened with similar settlement patterns in the past.[63]

These concerns have renewed the collaboration between indigenous peoples, riverside communities, research institutes, and environmental activists, especially international NGOs, and yielded a new network and a broader set of instruments for environmental action to prevent the construction of Belo Monte dam. These include traditional instruments, such as petitions, open letters, street protests, blockades and occupations, as well as more recent ones, such as technical (e.g., environmental impact assessments), legal (e.g., lawsuits and injunctions), political (e.g., the Ministry of the Environment), and mediatic (e.g., *Avatar* and Google Earth). Importantly though, the pressure point that was crucial to stop the project the first time around is gone this time: the dependence on foreign finance via the World Bank. This time, the Brazilian government is bankrolling the project through its National Bank for Economic and Social Development (BNDES), which will provide 80 percent of the financing.

The resistance to the project is led by the indigenous peoples of the Xingu, and especially the Kayapó, whose organizing skills were on display once again in the creation of the interethnic alliance of the "peoples of the Xingu" at the Piaraçu meeting of 2003,[64] and during the organization of the Second Meeting of the Peoples of the Xingu in Altamira in May 2008. The process has produced forms of "interethnic negotiation"[65] that have given indigenous politics a regional and glocal dimension that mirrors the emerging shape of environmental activism both globally and in Brazil.[66]

The principal network bringing together the different groups opposed to the Belo Monte dam is the Movement for the Development of the Transamazonian and Xingu (MDTX), established in 1991. The MDTX is the successor of the Movement for the Survival of the Transamazonian (MPST), created in the late 1980s. The new name illustrates the more forward-looking attitude toward the region, and it is partly this positive attitude that drives the actions of its members. This attitude is also illustrated in the name of the Xingu Alive Forever Movement (MXVPS), created in direct response to the Belo Monte dam. With well over 100 organizational affiliates across the region, the MDTX is the largest activist network in the north of Brazil.[67] The MDTX has garnered support to fight the Belo Monte dam from other networks and social movements, indigenous rights organizations, a handful of politicians, researchers from the several Brazilian universities, and several national and international NGOs.[68] The movement even has its own ecological martyr, Ademir Alfeu Federicci, known as Dema, killed in August 2001 when he was the head of the MDTX, and whose death is attributed in the region to his outspoken opposition to the dams.[69]

The movement has adapted its methods of resistance to the new context, using technical and legal strategies to challenge the viability and the legality of the project, and to expose the hidden agenda that lurks behind the current proposal. They refer to expert studies that indicate that the Belo Monte dam, operating alone, would be highly inefficient due to the seasonal variation in the water level of the Xingu River.[70] These studies conclude that the average annual electricity production would be roughly one-third of the installed capacity. This has massive implications for the project, because it gives credibility to the fears that if the Belo Monte dam goes ahead, it is very likely that some or all of the additional dams contemplated in the previous plans will follow. The Belo Monte dam would be the first of a network of dams.[71] This would significantly alter the social and environmental impact of the project by dramatically increasing the flooded area. Thus, whilst the Belo Monte dam by itself would not create a massive reservoir, the second-in-line, the Altamira (ex-Babaquara) dam situated upstream to regulate the flow would flood a massive 6,200 square kilometers.[72] If all six dams planned initially for the Xingu basin were to be built, the total flooded area would be 22,000 square kilometers.[73]

The activists have also used scientific studies to discredit the "clean energy" label attached to the project. Several studies have shown that due to the methane emissions from rotting vegetation, some tropical dams could emit more greenhouse gases than energy from fossil fuels.[74] The green alternative, they argue, is to invest in wind and solar technology, as well as in energy efficiency. The movement cites a WWF-Brazil study showing that, by 2020, Brazil could cut the expected demand for electricity by 40 percent through investments in energy efficiency.[75] The power saved would be equivalent to 14 Belo Monte plants.

The technical contestation of the project has sometimes worked in conjunction with the legal contestation. That is, expert studies, in conjunction with environmental legislation, have been used to launch legal suits against the project, providing some temporary relief in the form of injunctions. The legal avenue has also been used to stop the project on the basis of its ongoing unconstitutionality, alleging that the process has breached article 231, which demands that indigenous peoples affected by the dam be heard. In both cases, legal contestation has been facilitated by the office of the public ministry (*Ministério Público*), a body of independent public prosecutors charged with defending constitutional rights and public interests, created in 1992. The opponents to the dam have engaged the public ministry at different points of the process. The strategy has been temporarily successful on several occasions, mainly due to breaches of technical or procedural provisions, but once these were rectified, the courts have allowed the project to continue.[76] However, the decision about the constitutionality of the project is still pending, and activists are currently using this, as well as pointing out other financial, judicial and publicity risks associated with the Belo Monte dam, to pressure the financial institutions behind the project, including BNDES.[77]

The direct political avenue of contestation has proved so far to be of little help, despite the fact that the movement had a powerful ally inside the Brazilian government: Marina Silva, the minister of the environment. Since its creation in 1992, the ministry of the environment has paid special attention to the Amazon. However, it was the appointment of Marina Silva to this office in 2002 that was widely seen as the turning point in the recent history of Brazilian environmentalism, giving hope to the movement in general and in particular to those preoccupied with the future of the Amazon. She came into the ministry as a renowned environmental activist, a respected senator from the State of Acre, and a leader of the Peoples of the Forest movement created by Chico Mendes.[78] However, her presence in the cabinet betrayed the strong support for the developmentalist agenda at the heart of Lula's administration. Silva found herself in constant battle with many of her colleagues, reportedly without the support of the president, and often in direct confrontation with Dilma Rousseff, Lula's successor in the presidency. Increasingly isolated, Silva resigned in 2008 following a series of acrimonious disputes with fellow ministers and businessmen who accused her of stalling major development projects in the Amazon and hindering the Brazilian economy.[79] Chief amongst these disputes was her opposition to further hydroelectric dams in the Amazon.

The new minister of the environment, Green Party founder Carlos Minc, has not displayed the same passion for the Amazon—at one point having approved six new conservation units whilst having granted 152 new construction licenses. This was a stark imbalance considering the slogan he popularized: *dois pra lá, dois pra cá* (two that way, two this way), a commitment to approve environmental protection and developmental measures in equal measure.[80]

The political pressure on anyone opposing the Belo Monte dam from inside the administration has been intense and unrelenting. Administrators who have raised concerns about the project, including several senior officials such as the last two presidents of IBAMA, have resigned citing high-level political pressure to approve the project, that is, to grant licenses for the construction of the Belo Monte dam.[81] Others have been fired. One of the country's best-known indigenous leaders, the Kayapó Megaron Txucarramãe, was fired in November 2011 from his FUNAI position, reportedly because of his long-standing outspoken stance against the Belo Monte dam.[82]

In this context, the classic strategies of direct action and media campaigns have emerged as crucial to the potential success of the resistance to the Belo Monte dam. One of the most important actions of this kind has been the "Open Letter to the President," dated March 10, 2010, signed by Christian Poirier, Amazon Watch Brazil Program Coordinator, and endorsed by close to 150 organizations.[83] The letter urges the president to "immediately suspend" the project "due to the tremendous social, environmental, and economic risks" it poses to the Amazon region. The letter denounces the fact that the right of indigenous peoples to be consulted has not been honored, in breach of the Brazilian Constitution, the United Nations Declaration of the Rights of Indigenous Peoples (UNDRIP), and Convention 169 of the International Labour Organization (ILO). The letter cites the 2007 WWF-Brazil study proposing more efficient and clean alternatives to feed the energy needs of the country, and concludes that "Belo Monte is not only a bad predicament for the people of the Xingu, it is a bad investment for Brazil."

In response, Lula expressed his concern for the Amazon and for its indigenous peoples, but dismissed the letter as the work of foreign NGOs. This point was not lost on the indigenous peoples of the Xingu, who wrote a letter signed exclusively by indigenous leaders—62 of them—entitled "We, the indigenous people of the Xingu, do not want Belo Monte."[84] Notably, the letter stated directly: "We are not international NGOs." The signatories refer to the forest as their "butcher shop" and the river as their "market" and reiterate warnings that if the government proceeds with the dam, "they would have war on their hands." The letter also notes that the indigenous people of the Xingu had decided to invite the director of *Avatar,* James Cameron, to help them carry their message "to the entire world and to the Brazilians who do not yet know what is happening on the Xingu." Since then, the number of open letters, protests and petitions has continued to mount, including online petitions such as Amazon Watch's Stop the Belo Monte Monster Dam!, and official petitions and denunciations to the Inter-American Commission on Human Rights, in November 2010, the Organization of American States, in January 2011, and the United Nations Human Rights Council, in June 2011.

The government has faced off all these protests and is pushing ahead with the project.

In this context, what might emerge as the decisive strategy to stop the dam is the belligerent stance of the indigenous peoples of the Xingu, led by the Kayapó, coupled with their ability to utilize the media to present their case and stage their resistance. Indeed, at each stage of the project, the Kayapó have made references to "war" and "bloodshed," and these threats have increased as the construction phase of the Belo Monte dam approaches.[85] The most recent actions, the "Declaration against the Belo Monte dam"[86] and the occupation of the construction site in October 2011,[87] suggest that this might become an all-out war for the future of the Xingu, similar in nature, if not in methods, to that fought by the Na'vi in *Avatar.*

The invitation extended by the indigenous peoples of the Xingu to James Cameron and his team becomes particularly meaningful when placed in this context. *Avatar* is a sci-fi story of an indigenous people (the Na'vi) who fight an all-out war to defend their land and culture against outside forces bent on pillaging the resources of their land, the planet Pandora. The plot is as simplistic as the special effects are sophisticated, but the underlying dynamics at play in the film reflect rather well those at play in the Xingu region. Cameron himself has drawn a direct parallel between the plot of *Avatar* and the battle over the Belo Monte dam,[88] and has produced a 20-minute documentary, entitled *Message from Pandora,* that provides a full account of all the issues raised by the critics of the dam, with references to *Avatar,* and images from a Google Earth video of the Belo Monte dam.[89] Presented at Cannes in 2010, the documentary subsequently won the Cinema for Peace award, and has been included in the *Avatar* Extended Collector's Edition DVD.[90] The Google Earth video is a 10-minute production, created by Amazon Watch and International Rivers, and narrated by Sigourney Weaver, showing the impact of the Belo Monte dam on the Xingu and proposing greener alternatives.[91]

The involvement of foreign celebrities has brought renewed attention to the plight of the indigenous peoples of the Amazon, and their struggle to stop the Belo Monte dam.[92] But it has also fuelled one of the favorite strategies of the military and political elites: the portrayal of any foreign support for initiatives to protect the Amazon forest, and the human rights of Brazilian Indians, as expressions of Western imperialism, designed to take over the Amazon and impede the economic progress of Brazil.[93] Such cries of foreign imperialism conveniently overlook the fact that the same or very similar dynamics operate in the internal exploitation of the Amazon,[94] particularly when so-called national interests come into conflict with indigenous peoples, as is the case in the struggle over the Belo Monte dam.

The claims of imperialism also ignore the fact that there is plenty of opposition to the project and to the kind of development it represents amongst Brazilians—even amongst renowned film directors, such as Fernando Meirelles, who recently referred to the Belo Monte dam as "one of the most serious issues in Brazil today."[95] He went on to point out that the same mistakes of the past are being repeated here and urged Brazilians to discuss whether the progress they want for Brazil is that represented by the Belo Monte dam. The answer to this question will go a long way in shaping the Brazil of the 21st century.

* * * * *

The outcome of the struggle over the Belo Monte dam will have major repercussions for the future of the Amazon, but it will also determine what kind of nation Brazil will become. Will it be a nation of dams—hence the title of the chapter—or one sensitive to its natural environment, and respectful of the human rights of its indigenous peoples? The answer will partly depend on whether the government will continue to pursue a growth-at-all-cost developmental agenda or whether the environment—and the rights of indigenous peoples—will come to occupy a more central position in its plans for the future of the country. Given the current trajectory, there is little hope that the latter will happen any time soon without significant pressure on the Brazilian government.

Brazilian environmentalism will have to play a crucial role in the years ahead to transform the government's approach to the environment. The institutionalization of the concept of sustainable development presents a good opportunity to engage with the government and to keep sustainability, and thus the environment, as high in the political agenda as possible. The time Marina Silva spent as minister of the environment provided the best opportunity thus far to embed environmental concerns at the heart of the Brazilian government. The fact that she failed to succeed in that task is a clear indication of how much work lies ahead for environmental activists. Yet, the 19 percent of the vote she achieved in the 2010 presidential elections, as candidate for the Green Party, bodes well for the future of environmentalism in Brazil. Moreover, the environmental movement has now at its disposal a richness of resources, actors, and strategies unparalleled in Brazilian history. Thus, although the challenges ahead are huge, the potential for success should not be underestimated.

Some of the most important challenges environmentalism faces in the coming years are the endemic corruption and weak governance that affect all matters political in Brazil. In the first eight months of her presidency, Dilma Rousseff lost four ministers because of accusations of corruption—albeit as a result of her commitment to clean up Brazilian politics. The failures of the political system constitute a particular problem for a movement that increasingly tries to achieve environmental outcomes through direct engagement with government. If the state is weak and politicians are corrupt the ability to achieve meaningful change is severely diminished. The culture of political corruption and legal impunity is particularly troubling in the Amazon—a region that, rather than "a land without men," often resembles "a land without law." The murder of environmental activists over the years—including high-profile ones, such as Chico Mendes (1988), Dema (2001), Dorothy Stang (2005), and Zé Claudio (2011)—makes grassroots environmentalism in the Amazon a matter of life and death, and not just for the Amazon forest.

The political failings of the state, and the direct relevance these have for achieving environmental outcomes, suggest that environmental activists should reinstate the ecopolitical dimension as central to their engagement with the state and society in general, and include in their agenda for the future working to improve the quality of Brazilian democracy. They should also make central to their overall agenda the continuing engagement with indigenous peoples, especially regarding the future of the Amazon, and the role this region should play in 21st-century Brazil. Indigenous peoples, whilst not inherently conservationist, have the best record of maintaining natural ecosystems around them. Moreover, they have proven to be formidable political agents and there is no reason to believe they will not continue to be so. The alliance between environmental activists and indigenous peoples will be crucial to determine the outcome of the Belo Monte dam, and this outcome will go a long way in determining what direction Brazil will take in the 21st century.

Acknowledgments

The author wishes to thank Adela Alfonsi for her editorial assistance with the chapter, and Sandra Elhelw for her research assistance in the early stages of the project.

Notes

1. Heckenberger et al., "Cultural Landscapes," 107.
2. Bentes, "Intervenção do ambientalismo internacional."
3. Denevan, "The Pristine Myth."
4. Heckenberger et al., "Amazonia 1492."
5. Bolaños, "Redefining Identities."
6. Gibbons, "New View of Early Amazonia," and Baleé, "Natives Views."
7. Ramos, *Indigenism.*
8. Hemming, *Tree of Rivers.*
9. Hemming, *Red Gold.*
10. Diacon, *Stringing Together a Nation.*
11. Andrade, "Conquistar a Terra."
12. Lima, *Um grande cerco de paz.*
13. Fisher, "Megadevelopment, Environmentalism, and Resistance," 224.
14. Fearnside, "Deforestation in Brazilian Amazonia," 680.
15. Franco and Drummond, "Wilderness and the Brazilian Mind (II)."
16. Pádua, "O nascimento da política verde no Brasil."
17. Alonso, Costa, and Maciel, "Environmental Activism in Brazil," 137.
18. Oliveira, "Gênese e redefinições," 756–57.
19. Bentes, "Intervenção do ambientalismo internacional."
20. Le Tourneau and Bursztyn, "Assentamentos rurais na Amazônia."
21. Fisher, "Megadevelopment, Environmentalism, and Resistance," 224.
22. Schwartzman and Zimmerman, "Conservation Alliances," 722.
23. Worcman, "Brazil's Thriving Environmental Movement."
24. Alonso, Costa, and Maciel, "Environmental Activism in Brazil."
25. Loureiro and Pacheco, "Formação e consolidação."
26. Alonso, Costa, and Maciel, "Environmental Activism in Brazil."
27. Oliveira, "Gênese e redefinições," and Alonso, Costa and Maciel, "Environmental Activism in Brazil."
28. Hochstetler and Keck, *Greening Brazil,* 158.
29. Hochstetler and Keck, *Greening Brazil,* 158–59.
30. Pádua, "O nascimento da política verde no Brasil."
31. Zhouri, "Transnational Campaigns for the Amazon."
32. Viola, "The Ecologist Movement in Brazil," and Jacobs, "Community Participation."
33. Hochstetler and Keck, *Greening Brazil.*
34. Keck and Sikkink, *Activists beyond Borders.*
35. Keck, "Dilemmas for Conservation," 38.
36. López, "Environmental Change," and da Costa, "Brazil's SIVAM."
37. Landim, *A invenção das ONGs.*
38. Alonso, Costa, and Maciel, "Environmental Activism in Brazil."
39. Tesh and Paes-Machado, "Sewers, Garbage, and Environmentalism."
40. Alonso, Costa, and Maciel, "Environmental Activism in Brazil."
41. Alonso and Maciel, "From Protest to Professionalization."

42. Alonso and Maciel, "From Protest to Professionalization."
43. Carneiro, "A construção social."
44. Zhouri, "Global-Local Amazon Politics."
45. Carvalho, "Environmental Resistance," 257.
46. Carvalho, "Environmental Resistance," 257.
47. Fearnside, "Dams in the Amazon," 20.
48. Diamond and Poirier, "Brazil's Native Peoples."
49. Ramos, *Indigenism,* 217.
50. Fisher, "Megadevelopment, Environmentalism, and Resistance," 222.
51. Fisher, "Megadevelopment, Environmentalism, and Resistance," 222.
52. Fearnside, "Dams in the Amazon," 16.
53. Fisher, "Megadevelopment, Environmentalism, and Resistance," 223.
54. Turner, "Defiant Images."
55. Fisher, "Megadevelopment, Environmentalism, and Resistance," 222.
56. Carvalho, "Environmental Resistance," 245.
57. Viola, "Emergency and Development," 322.
58. Schaeffer et al., "Indicators for Sustainable Energy Development."
59. Fearnside, "Dams in the Amazon," 21.
60. da Fonseca and Bourgiognie, "The Belo Monte Dam Case," 104.
61. Diamond and Poirier, "Brazil's Native Peoples."
62. Anaya, "Report by the Special Rapporteur."
63. Carvalho, "Hydroelectric Development," and Le Tourneau and Bursztyn, "Assentamentos rurais na Amazônia."
64. Turner and Fajans-Turner, "Political Innovation."
65. Garnelo and Sampaio, "Globalização e ambientalismo," 766.
66. Pieck, "Opportunities."
67. McCormick, "The Brazilian Anti-dam Movement."
68. Fearnside, "Dams in the Amazon," 19, and McCormick, "Damming the Amazon."
69. Fearnside, "Dams in the Amazon," 22.
70. da Fonseca and Bourgiognie, "The Belo Monte Dam Case," 105.
71. McCormick, "Damming the Amazon," 41.
72. Fearnside, "Dams in the Amazon."
73. Carvalho, "Environmental Resistance," 257.
74. Fearnside, "Environmental Impacts."
75. WWF-Brazil, "WWF-Brazil's Sustainable Power."
76. Scholz, "Environmental Policy," and Carvalho, "Environmental Resistance."
77. Xingu Vivo, "Bancos recebem notificação."
78. Rodrigues, *Walking the Forest with Chico Mendes.*
79. Barrionuevo, "'Stagnation' made Brazil's Environment Chief Resign."
80. Angelo, "PT tenta apagar fama 'antiverde' de Dilma."
81. da Fonseca and Bourgiognie, "The Belo Monte Dam Case," 104.
82. Txucarramãe, "Letter from Indigenous Leader," and Phillips, "Brazilian Anti-dam Campaigner Sacked."
83. Poirier, "Open Letter to President Luiz Inácio Lula da Silva."

84. Kayapó and Juruna, "We, the Indigenous People of the Xingú."

85. See, for example, Amato, "Indios ameaçam 'Guerra'," and Phillips, "Awarding of Brazilian Dam Contract Prompts Warning of Bloodshed."

86. Xingu Alliance, "Declaration against the Belo Monte."

87. Sommerfilms, "Indigenous Peoples Historic Occupation of Belo Monte," and Amazon Watch, "Hundreds Occupy Belo Monte Dam Site in Brazilian Amazon."

88. Phillips, "Avatar Director James Cameron Joins Amazon Tribe's Fight."

89. Amazon Watch, "James Cameron's 'A Message from Pandora.'"

90. Amazon Watch, "A Message from Pandora wins Peace Award."

91. Amazon Watch, "Defending the Rivers of the Amazon, with Sigourney Weaver."

92. See, for example, Yapp, "Brazil's Belo Monte Dam."

93. Zhouri, "Adverse Forces in the Brazilian Amazon."

94. Bentes, "Intervenção do ambientalismo internacional."

95. Xingu Vivo, "Diretores do filme 'Xingu' criticam Belo Monte em Manaus."

Bibliography

Alonso, A., V. Costa, and D. Maciel. "Environmental Activism in Brazil: The Rise of a Social Movement." In *Citizenship and Social Movements: Perspectives from the Global South,* edited by L. Thompson and C. Tapscott, 131–59. London: Zed Books, 2010.

Alonso, A., and D. Maciel. "From Protest to Professionalization: Brazilian Environmental Activism after Rio-92." *Journal of Environment & Development* 19, no. 3 (2010): 300–17.

Amato, F. "Indios ameaçam 'Guerra' contra construção da usina Belo Monte." *Folha de São Paulo,* June 2, 2010. http://www1.folha.uol.com.br/mercado/744918-indios-ameacam-guerra-contra-construcao-da-usina-belo-monte.shtml (Accessed August 27, 2011).

Amazon Watch. "Defending the Rivers of the Amazon, with Sigourney Weaver." August 30, 2010. http://amazonwatch.org/news/2010/0830-defending-the-rivers-of-the-amazon-sigourney (Accessed September 5, 2011).

Amazon Watch. "Hundreds Occupy Belo Monte Dam Site in Brazilian Amazon." October 27, 2011. http://amazonwatch.org/news/2011/1027-hundreds-occupy-belo-monte-dam-site-in-brazilian-amazon (Accessed October 30, 2011).

Amazon Watch. "James Cameron's film A Message from Pandora wins International Cinema for Peace Award." February 15, 2011. http://amazonwatch.org/news/2011/0215-james-camerons-film-a-message-from-pandora-wins-international-cinema-for-peace-award (Accessed September 5, 2011).

Amazon Watch. "Just Released: James Cameron's 'A Message from Pandora,' a short-film about a Real Avatar Battle in the Amazon Rainforest." November 17, 2010. http://amazonwatch.org/news/2010/1117-just-released-james-camerons-a-message-from-pandora (Accessed September 5, 2011).

Anaya, J. "Report by the Special Rapporteur on the Situation of Human Rights and Fundamental Freedoms of Indigenous People." Human Rights Council. A/HRC/15/37/add.1 (2010). http://unsr.jamesanaya.org/PDFs/Communications%20report-FINAL.pdf (Accessed September 5, 2011).

Andrade, R. de P. "'Conquistar a terra, dominar a água, sujeitar a floresta': Getúlio Vargas e a revista 'Cultural Política' redescobrem a Amazônia (1940–1941)." *Boletim do Museu Paraenense Emílio Goeldi. Ciências Humanas* 5, no. 2 (2010): 453–68.

Angelo, C. "PT tenta apagar fama 'antiverde' de Dilma." *Folha de São Paulo,* October 10, 2010. http://www1.folha.uol.com.br/poder/812470-pt-tenta-apagar-fama-antiverde-de-dilma.shtml (Accessed August 27, 2011).

Baleé, W. "Natives Views of Environment in Amazonia." In *Nature across Culture: Views of Nature and Environment in non-Western cultures,* edited by S. Helaine, 277–88. Boston: Kluwer Academic Publishers, 2003.

Barrionuevo, A. "'Stagnation' made Brazil's Environment Chief Resign." *The New York Times,* May 16, 2008. http://www.nytimes.com/2008/05/16/world/americas/16brazil.html?pagewanted=print (Accessed September 5, 2011).

Bentes, R. "A intervenção do ambientalismo internacional na Amazônia." *Estudos Avançados* 19 (2005): 225–40.

Bolaños, O. "Redefining Identities, Redefining Landscapes: Indigenous Identity and Land Rights Struggles in the Brazilian Amazon." *Journal of Cultural Geography* 28, no. 1 (2011): 45–72.

Carneiro, M.S. "A construção social do Mercado de madeiras certificadas na Amazônia brasileira: A Atuação das ONGs ambientalistas e das empresas pioneiras." *Sociedade e Estado* 22, no. 3 (2007): 681–713.

Carvalho, G.O. "Environmental Resistance and the Politics of Energy Development in the Brazilian Amazon." *Journal of Environment & Development* 15, no. 3 (2006): 245–68.

Carvalho, G.O. "Hydroelectric Development and Road Paving in Brazil's Transamazon area." *Journal of Environment & Development* 8, no. 4 (1999): 397–406.

da Costa, T.G. "Brazil's SIVAM: As it Monitors the Amazon, Will It Fulfill its Human Security Promise?" ECSP Report 7 (2001): 47–58. http://www.wilsoncenter.org/publication/brazils-sivam-it-monitors-the-amazon-will-it-fulfill-its-human-security-promise (Accessed August 23, 2011).

da Fonseca, P.G., and A. Bourgiognie. "The Belo Monte Dam Case." *Environmental Policy and Law* 41, no. 2 (2011): 104–7.

Denevan, W.M. "The Pristine Myth: The Landscape of the Americas in 1492." *Annals of the Association of American Geographers* 82, no. 3 (1992): 369–85.

Diacon, T.A. *Stringing Together a Nation: Cândido Mariano da Silva Rondon and the Construction of a Modern Brazil, 1960–1930.* Durham: Duke University Press, 2004.

Diamond, S., and C. Poirier. "Brazil's Native Peoples and the Belo Monte Dam: A Case Study." *NACLA Report on the Americas* 43, no. 5 (2010): 25–29.

Fearnside, P.M. "Avança Brasil: Environmental and Social Consequences of Brazil's Planned Infrastructure in Amazonia." *Environmental Management* 30, no. 6 (2002): 735–47.

Fearnside, P.M. "Dams in the Amazon: Belo Monte and Brazil's Hydroelectric Development of the Xingu River Basin." *Environmental Management* 38, no. 1 (2006): 16–27.

Fearnside, P.M. "Deforestation in Brazilian Amazonia: History, Rates, and Consequences." *Conservation Biology* 19, no. 3 (2005): 680–88.

Fearnside, P.M. "Environmental Impacts of Brazil's Tucuruí Dam: Unlearned Lessons for Hydroelectric Development in Amazonia." *Environmental Management* 27, no. 3 (2001): 377–96.

Fisher, W.H. "Megadevelopment, Environmentalism, and Resistance: The Institutional Context of Kayapó Indigenous Politics in Central Brazil." *Human Organization* 53, no. 3 (1994): 220–32.

Franco, J. L de A., and J.A. Drummond. "Wilderness and the Brazilian Mind (II): The First Brazilian Conference on Nature Protection (Rio de Janeiro, 1934)." *Environmental History* 14, no. 1 (2009): 82–102.

Garnelo, L., and S. Sampaio. "Globalização e ambientalismo: etnicidades polifônicas na Amazônia." *História, Ciências, Saúde* 12, no. 3 (2005): 755–68.

Gibbons, A. "New View of Early Amazonia." *Science* 248 (1990): 1488–90.

Heckenberger, M.J., A. Kuikuro, U.T. Kuikuro, J.C. Russell, M. Schmidt, C. Fausto, and B. Franchetto. "Amazonia 1492: Pristine Forest or Cultural Parkland?" *Science* 301 (2003): 1710–14.

Heckenberger, M.J., J.C. Russell, J.R. Toney, and M.J. Schmidt. "The Legacy of Cultural Landscapes in the Brazilian Amazon: Implications for Biodiversity." *Philosophical Transactions of the Royal Society* 362 (2007): 197–208.

Hemming, J. *Red Gold: The Conquest of the Brazilian Indians.* London: Macmillan, 1978.

Hemming, J. *Tree of Rivers: The Story of the Amazon.* London: Thames & Hudson, 2008.

Hochstetler, K., and M.E. Keck. *Greening Brazil: Environmental Activism in State and Society.* Durham: Duke University Press, 2007.

Jacobs, J.E. "Community Participation, the Environment, and Democracy: Brazil in Comparative Perspective." *Latin American Politics and Society* 44, no. 4 (2002): 59–88.

Kayapó, C.B.K., C.R. Kayapó, and Y. Juruna. "We, the indigenous people of the Xingú, do not want Belo Monte." April 20, 2010. http://assets.survival-international.org/documents/264/We_the_indigenous_people_of_the_Xing_do_not_want_Belo_Monte.pdf (Accessed February 10, 2011).

Keck, M.E. "Dilemmas for Conservation in the Brazilian Amazon." ECSP Report 7 (2001): 32–46. http://www.wilsoncenter.org/publication/dilemmas-for-conservation-the-brazilian-amazon (Accessed August 23, 2011).

Keck, M.E., and K. Sikkink. *Activists beyond Borders: Advocacy Networks in International Politics.* Ithaca: Cornell University Press, 1998.

Landim, L. *A invenção das ONGs: Do servico invisível à profissão sem nome.* Rio de Janeiro: Museu Nacional/UFRJ, 1993.

Le Tourneau, F.M., and M. Bursztyn. "Assentamentos rurais na Amazônia: Contradições entre a política agraria e a política ambiental." *Ambiente & Sociedade* 13, no. 1 (2010): 111–30.

Lima, A. C. de S. *Um grande cerco de paz: Poder tutelar, indianidade e formação do Estado no Brasil.* Petrópolis: Vozes, 1995.

López, A. "Environmental Change, Security, and Social Conflicts in the Brazilian Amazon." *Environmental Change & Security Project Report* 5 (1999): 26–33.

Loureiro, M.R., and R.S. Pacheco. "Formação e consolidação do campo ambiental no Brasil: Consensos e disputas (1972–1992)." *Revista de Administração Pública* 29, no. 4 (1995): 137–53.

McCormick, S. "The Brazilian Anti-dam Movement: Knowledge Contestation as Communicative Action." *Organization & Environment* 19, no. 3 (2006): 321–46.

McCormick, S. "Damming the Amazon: Local Movements and Transnational Struggles over Water." *Society and Natural Resources* 24, no. 1 (2011): 34–48.

Oliveira, W.J.F. "Gênese e redefinições do militantismo ambientalista no Brasil." *Dados* 51, no. 3 (2008): 751–77.

Pádua, J.A. "O nascimento da política verde no Brasil: Fatores endógenos e exógenos." In *Ecologia e política mundial,* edited by H. Leis, 135–62. Rio de Janeiro: FASE/Vozes/AIRI/PUC-Rio, 1991.

Phillips, T. "Avatar Director James Cameron Joins Amazon Tribe's Fight to Halt Giant Dam." *The Guardian,* April 18, 2010. http://www.guardian.co.uk/world/2010/apr/18/avatar-james-cameron-brazil-dam/print (Accessed December 20, 2010).

Phillips, T. "Awarding of Brazilian Dam Contract Prompts Warning of Bloodshed." *The Guardian,* April 21, 2010. http://www.guardian.co.uk/environment/2010/apr/21/contract-belo-monte-dam/print (Accessed December 20, 2010).

Phillips, T. "Brazilian Anti-dam Campaigner Sacked." *The Guardian,* November 2, 2011. http://www.guardian.co.uk/world/2011/nov/02/brazil-anti-dam-campaigner-sacked (Accessed November 4, 2011).

Pieck, S.K. "Opportunities for Transnational Indigenous Eco-politics: The Changing Landscape in the New Millennium." *Global Networks* 6, no. 3 (2006): 309–29.

Poirier, C. "Open Letter to President Luiz Inácio Lula da Silva." March 10, 2010. http://assets.survivalinternational.org/documents/269/Appeal_letter_English.pdf (Accessed February 10, 2011).

Ramos, A. *Indigenism: Ethnic Politics in Brazil.* Madison: The University of Wisconsin Press, 1998.

Rodrigues, G. *Walking the Forest with Chico Mendes: Struggle for Justice in the Amazon.* Edited and translated by Linda Rabben. Austin: The University of Texas Press, 2007.

Schaeffer, R., A. Szklo, F. Cima, and G. Machado. "Indicators for Sustainable Energy Development: Brazil's Case Study." *Natural Resources Forum* 29, no. 4 (2005): 284–97.

Scholz, I. "Environmental Policy Cooperation among Organized Civil Society, National Public Actors and International Actors in the Brazilian Amazon." *European Journal of Development Research* 17, no. 4 (2005): 681–705.

Schwartzman, S., and B. Zimmerman. "Conservation Alliances with Indigenous Peoples of the Amazon." *Conservation Biology* 19, no. 3 (2005): 721–27.

Sommerfilms. "Indigenous Peoples Historic Occupation of Belo Monte." YouTube video. October 27, 2011. http://www.youtube.com/watch?v=USXXO8NoNss (Accessed October 30, 2011).

Tesh, S. N., and E. Paes-Machado. "Sewers, Garbage, and Environmentalism in Brazil." *Journal of Environment & Development* 13, no. 1 (2004): 42–72.

Turner, T. "Defiant Images: The Kayapo Appropriation of Video." *Anthropology Today* 8, no. 6 (1992): 5–16.

Turner, T., and V. Fajans-Turner. "Political Innovation and Inter-ethnic Alliance: Kayapo Resistance to the Developmentalist State." *Anthropology Today* 22, no. 5 (2006): 3–10.

Txucarramãe, M. "Letter from Indigenous Leader Megaron Txucarramãe to the International Press." April 26, 2010. http://assets.survivalinternational.org/documents/265/Letter_from_indigenous_leader_Megaron_Tx_ucarram_e_to_the_International_Press.pdf (Accessed February 10, 2011).

Viola, E. J. "The Ecologist Movement in Brazil (1974–1986): From Environmentalism to Ecopolitics." *International Journal of Urban and Regional Research* 12, no. 2 (1998): 211–28.

Viola, E. J. "Emergency and Development of the International Regime of Climate Change (1990–2005)." In *A Contribution to Understanding of Global Change in South America,* edited by P.L. da S. Dias, W.C. Ribeiro, and L.H. Nunes, 313–29. São Paulo: Instituto de Estudos Avançados da Universidade de São Paulo, 2007.

Worcman, N. B. "Brazil's Thriving Environmental Movement." *Technology Review* 93, no. 7 (1990): 42–51.

WWF-Brazil. "WWF-Brazil's Sustainable Power Sector Vision 2020." 2007. http://www.climatesolver.org/source.php?id=1252339 (Accessed September 5, 2011).

Xingu Alliance. "Declaration against the Belo Monte Dam." October 26, 2011. http://earthpeoples.org/blog/?p=799 (Accessed November 6, 2011).

Xingu Vivo. "Bancos recebem notificação sobre riscos de envolvimento com Belo Monte." November 7, 2011. http://www.xinguvivo.org.br/2011/11/07/bancos-recebem-notificacao-sobre-riscos-de-envolvimento-com-belo-monte/ (Accessed November 8, 2011).

Xingu Vivo. "Diretores do filme 'Xingu' criticam Belo Monte em Manaus." November 5, 2011. http://www.xinguvivo.org.br/2011/11/05/1681/ (Accessed November 8, 2011).

Yapp, R. "Brazil's Belo Monte Dam: A History of Celebrity Opposition." *The Telegraph,* January 28, 2011. http://www.telegraph.co.uk/news/worldnews/

southamerica/brazil/8287289/Brazils-Belo-Monte-dam-a-history-of-celebrity-opposition.html (Accessed September 5, 2011).

Zhouri, A. "'Adverse forces' in the Brazilian Amazon: Developmentalism versus Environmentalism and Indigenous Rights." *Journal of Environment & Development* 19, no. 3 (2010): 252–73.

Zhouri, A. "Global-Local Amazon Politics: Conflicting Paradigms in the Rainforest Campaign." *Theory, Culture & Society* 21, no. 2 (2004): 69–89.

Zhouri, A. "Transnational Campaigns for the Amazon: NGO Strategies, Trade and Official Responses." *Ambiente & Sociedade* 6/7 (2001): 31–63.

Chapter 4

Neo-extractivism, Neo-rentierism, and Social Movements in Venezuela's 21st-Century Ecosocialism: Discourses, Conflicts, and Resistance

María Pilar García-Guadilla

> It is becoming clear that for the progressive governments or for the new left governments, environmental issues have become a flank of serious contradictions. Their strong support to extractive activities to fuel economic growth is aggravating environmental impacts, triggers serious social protests, and perpetuates the subordination to be suppliers of raw materials for globalization.
>
> —Eduardo Gudynas[1]

Introduction: New Questions about the Old New Left Governments' Discourses

In response to the mobilization of socio-environmental movements, Latin American governments incorporated the discourses of sustainable development and participatory democracy into regional agendas such as the Latin American and Caribbean Commission on Development and the Environment's "Our Own Agenda" of 1990. They concurred with the official Rio Declaration of 1992 on the discourse on sustainable development expressed in the 1987 World Commission on the Environment and Development's Brundtlandt Report, but they differed on the relevant role they assigned to democracy; they emphasized the need to go beyond representative democracy and to complement it with "participatory democracy." In the 1992 Rio de Janeiro Summit, Latin American governments and social movements pointed to extensive poverty and underdevelopment as environmental problems per se, emphasizing that to achieve sustainable development, it was first necessary to decrease poverty and to increase international and national social equity. They also shared the critique with many governments and social movements of developing countries (the G77 in particular) that economic globalization and its corollaries, free trade and privatization, exacerbate socio-environmental problems instead of diminishing them, as industrialized countries and international institutions argued they would. Despite their criticism of capitalism-neoliberal-globalizing development, Latin American governments did not propose an alternative in the 1992 Rio Summit and for almost a decade, environmental policy sharply departed from their discourses and "good intentions." In contrast to their governments' official position, the Alternative Agenda that social organizations and movements signed in Rio de Janeiro in 1992 contained a new rationale and paradigm to achieve sustainable development through participatory democracy. There were not many changes between the 1992 World Summit in Rio de Janeiro and the one in Johannesburg in 2002 with regard to the Latin American governments' conceptions of "sustainable development" and "participatory democracy"; the same was true for social movements, with both parties continuing to support their 1992 agenda.[2]

During the decade 2002–2012, new concepts were incorporated into the debate, such as the "green economy" originally introduced at the 2002 Johannesburg Summit by multilateral international organizations, and new notions of citizenship based upon indigenous and nature's rights and ancestral conceptions such as the *Buen Vivir* (Good Living), the *Sumak Kawsay*, and the *Pachamama* proposed by the indigenous communities and the socio-environmental movements.[3] Currently, Latin American governments are divided into two blocs: those who support the multilateral international organizations' conception of sustainable development expressed in Agenda 21 and the proposal for a green economy, and the New Leftist governments such as the Venezuelan, Bolivian, and Ecuadorian governments whose discourses in environmental summits are closer to the Alternative Agenda of social movements, since they are in favor of transforming the capitalist neoliberal model but, paradoxically, they subscribe the green economy proposal and continue reinforcing the extractivist development model.

Within this context, the present study attempts to interpret current conflicts between the Chávez government and the socio-environmental movement and answer the following questions: to what extent do Venezuela's national and regional development plans offer a green agenda to comply with its anticapitalist, antineoliberal, antiimperialist, and antiglobalization discourse? Can an extractive economy such as Venezuela's become sustainable in the context of the self-styled 21st-century socialism, the highly political and ideological polarization of the environmental and indigenous issues, and the negative environmental and sociocultural impacts that derive from the extractive productive model? Do domestic and regional development plans, policies, and projects meet the demands of the socio-environmental movements? Is it possible to build through participatory democracy the alternative to the extractive economy proposed by the environmental movement that protects the environment and complies with the new constitutional rights of the indigenous communities within the socialism of the 21st century?

Venezuelan Neo-extractivism, Neo-rentierism, and Neocolonialism

Venezuela has an exporting mineral-extractive economy that is based on the exploitation of nonrenewable natural resources. At the end of the 1920s, the country became an exporter of oil; since then, oil income has been used to open up the doors to economic growth and democracy. Oil dependency has increased, representing 95 percent of the total value of exports for the year 2010.[4]

After the fall of the Pérez Jiménez dictatorship in 1958, representative-liberal democracy was incorporated into the 1961 Constitution and oil income was used to neutralize conflicts among the elites. From the 1960s to the 1990s, oil income accelerated the unequal pattern of urbanization and economic growth, to the extent that it was primarily appropriated by the emerging upper and middle classes; peasants and rural migrants who were displaced by the changes of the economy from agriculture to oil had difficulties getting incorporated into the modern, industrial, and urban sectors of the economy. As a result, poverty and social inequalities increased.

Based on the theory of the growth poles,[5] Venezuelan industrialization of the 1960s and 1970s was based on the exploitation and transformation of natural resources. In an attempt to diversify the economy beyond oil, at the beginning of the 1960s, the Guayana Industrial Project in the Bolívar State was initiated with the construction of large industrial complexes such as the steel mill and the aluminum plants and other infrastructure to transform and industrialize existing resources.[6] In this project, environmental considerations were subject to the achievement of greater economic efficiency, productivity, and competitiveness, creating an enclave that led to environmental and higher social costs for the population of Ciudad Guayana than any other nonplanned city in the country.[7] In Venezuela, the economic crisis of the 1980s and declining oil prices affected oil revenues and thus, the continuation of investments and economic growth based on large industrial projects and infrastructure; it contributed also to the loss of political legitimacy of rulers who did not have the resources for loosening the acute political tensions that existed at the time. Thus, the 1980s and 1990s were characterized by an acute political and economic crisis.

President Chávez, a self-defined leftist, took office in 1999 as oil prices started to increase. Despite his denunciation of capitalism and neoliberalism, rising oil prices led the government to expand the export productive model based on oil and mining projects, as part of his Socialism of the 21st Century. This orientation was translated into development plans and into policies and projects that continue reproducing the oil-rent economy and the extractive-productive model and have generated strong socio-environmental impacts and hence, conflicts, between the state and the indigenous, environmental, and human rights social movements (hereafter called socio-called socio-environmental movement).

The most emblematic environmental and indigenous conflicts of the Chávez era come from previous governments: the installation of an electric high-voltage wiring that crossed the Canaima National Park and affected the rich biodiversity and the indigenous habitats;[8] the approval for mining in the ecologically sensitive Imataca Forest Reserve;[9] the exploitation of gold and diamonds in the Caroní river basin, a tributary of the Orinoco River; the increase of exploitation of coal in the Sierra de Perijá, which is the main habitat for the Yutpas, the Bari, and other indigenous groups in the Zulia state, frontier with Colombia;[10] and the South American mega-infrastructures called the "Initiative for the Integration of Regional South America Infrastructure (IIRSA)," proposed, among other objectives, for the exportation of hydrocarbons in order to achieve sovereignty and to integrate energetically the Latin American and Caribbean countries. These conflicts have been located in ecologically sensitive areas that also correspond to the habitat of the indigenous communities; for this reason, it is not possible to separate the environmental struggles of the indigenous population from the constitutional claims to their territory and culture.

The new Constitution of 1999, derived from a constituent process, incorporated many of the socio-environmental movements' demands concerning environmental, indigenous, gender, socioeconomic, and human rights, and these constitutional visions of endogenous-sustainable development and participatory democracy were incorporated into the Social and Economic National Plan of Development 2001–2007 (SENPD) and the First Socialist Plan of Social and Economic Development 2007–2013 (FSPSED).[11] Nonetheless, in these national plans, development continued to be tied to the exploitation of oil and other nonrenewable natural resources and the scope of oil was expanded geopolitically as a tool for the Caribbean and South America integration and sovereignty. The slogans of the PDVSA, Venezuela's national oil company, were Full Sovereignty, Energy Union, and to rescue the oil sovereignty of PDVSA from transnational interests through the "internalization of hydrocarbons."[12] According to the SENPD and the FSPSED's development national plans, economic growth was to be based on intensive use of natural resources, mostly hydrocarbons, and had the goal of activating human, social, and local-ethnic development, improving the quality of life of the population, and subsidizing the distributive policies or social missions (Misiones Sociales)[13] to achieve social and political legitimacy.

The government environmental discourse has argued that hydrocarbons play a key revolutionary role[14] "in the fight against poverty for national and international sustainable development" and that oil income becomes essential "to finance social, educational, health and economic programs, and the productive missions coordinated by the respective ministries."[15] Socialism of the 21st Century considers oil income an indispensable resource to ensure national development, social development, equity and sovereignty; implicitly, it is considered a means to greater governance and social and political legitimacy. This "revolutionary-social distributive" role led to a new form of rentierism that could be named as "neo-rentierism," and it is implicit in PDVSA's policies that have the role of stimulating the social missions. PDVSA's contribution to the Fund for Economic and Social Development of the country (FONDESPA), which promotes the integral development of Venezuelans using the "revolutionary" oil income and by giving support for social missions, "is one of the ways in which PDVSA became involved with all Venezuelans, aspiring to a comprehensive human development within the framework of the principles laid down in the Constitution of the Bolivarian Republic of Venezuela."[16] The government has defined these social missions as well as the financing policies addressed to the extensive network of Communal Councils[17] as tools for inclusion.

In addition to being a "revolutionary" tool for inclusion and a tool against imperialism and globalization, oil, gas and other hydrocarbons in Venezuela are also considered essential mechanisms for economic, commercial, political, and physical-infrastructural integration[18] and implicitly a weapon for wanted hegemony in the Caribbean region through PETROCARIBE and through multiple solidarity and social-distributive policies displayed within the Caribbean Community (CARICOM). Most of the subregional and regional integration institutions such as CAN (Comunidad Andina de Naciones), MERCOSUR (Mercado Común del Sur), ALBA (Alianza Bolivariana para los Pueblos de Nuestra América), UNASUR (Unión de Naciones Suramericanas), and CARICOM conceived integration as a first step to enter the global economy;[19] they emphasize the need for regional trade liberalization and the complementation of economic and commercial agreements with physical integration through what has been called the Initiative for the Integration of the Regional Infrastructure of South America (IIRSA).[20] The IIRSA was created in the year 2000 and contemplates the consolidation of an important network of communications through the creation of terrestrial, fluvial, and transportation infrastructure that will allow the efficient movement of goods, information, and energy;[21] it will be mainly financed by regional institutions such as Inter-American Development Bank (IDB) and the Corporación Andina de Fomento (CAF), and cofinanced by each South American country and by other entities such as the European Union. Thus, the IIRSA is one of the main mechanisms contemplated by the variety of regional institutions for greater competitiveness in the exploitation of natural resources and in the production and transportation of goods and services, and to stimulate economic growth through expanding the productive area to national, regional, and global markets. It involves 507 projects in transport infrastructure, energy, and telecommunications distributed in 10 axes of development in which Venezuela participates in two: the Andean axis with Colombia, Ecuador, Bolivia, and Peru and in the Guiana axis with Brazil, Guyana, and Suriname.[22] After more than a decade of its creation in the year 2000, the IIRSA has been retaken by the Union of South American Nations (UNASUR) under the name of COSIPLAN (Consejo Suramericano de Infraestructura y Planeamiento).[23]

From National and Regional Plans to Reality: Productivist-Developmental Logic, Contradictions, and Conflicts

According to the SENPD and the FSPSED national development plans, economic development was supposed to emphasize the exploitation of natural resources under the government's assumption that such exploitation can be environmentally sustained and that mining could be "competitive and ecological" at the same time. The SENPD states that "to achieve sustainable economic development and to increase mining production and the competitiveness of exports in international markets, mining should be managed rationally"; it also points out the need to preserve the environment through the monitoring of the mining environmental impacts in order to improve the economic and social conditions of the regions.[24]

This assumption that implies the subordination of the socio-environmental dimension did not hold due to the magnitude and extent of the ecological and sociocultural negative impacts. Moreover, the attempt to greenwash the extractive economy through the concept of "ecological mining" that is used in the SENPD looks environmentally unviable since many extractive economic productive activities are located in fragile biodiverse areas and in the indigenous territories. Those activities contradict the constitutional mandate that the promotion of such activities must be performed by the state in accordance with the ethno-cultural characteristics of each community. Moreover, the government's arguments did not foresee the potential conflicts that could emerge as the result of the contradictory objectives and goals involved in the national development plans; thus, no guidelines were included to resolve those conflicts.

Another potential conflict that derives from development plans is how to combine economic and social interests, given the inclusion in the plans of "the social economy" that should be developed by the state to support its distributive social policies, and it is linked to the notion of social justice; this notion of social justice was aimed at democratizing the market and the economy through the participation of local communities and workers in the so-called alternative enterprises. The negative impacts of proposed economic policies and projects does not guarantee achieving the social objectives unless the interrelated aims of sustainability are balanced versus social equity; otherwise it will be difficult, if not impossible, to achieve both aims.

One aim of the FSPSED was "curbing the growth rate," provided that this did not affect the development of the extensive network of social policies, such as the social missions; nonetheless, given the close link between the oil revenues and the extensive social programs, the possibility of reducing the exploitation of hydrocarbons looks unviable since this is occurring while current oil prices are high; the probability of curbing growth in a scenario of decreasing oil prices will be even lower. Economic diversification was also included as an objective of the development plans, particularly the SENPD, but it was not translated into concrete policies and projects.

At least two of the seven general guidelines included in the SENPD, such as the "New National Geopolitics" and "Venezuela as a World Energy Power," could come in contradiction. While the New National Geopolitics stresses the need to protect certain sensitive areas to conserve water and biodiversity, preserve the balance of biodiversity-rich ecosystems, achieve an environmentally sustainable model of production and accumulation, and reduce the environmental impact of human intervention to become a World Energy Power, it would be required to increase the production of hydrocarbons and to build mega–infrastructures that will affect the ecosystems in the exploitation areas where there is high biodiversity.

Aside from the gap between the plans and the policy implementation, other conflicts have to do with the government's refusal to visualize the pernicious effects of the extractive economy due to the idea promoted by President Chávez and previous governments through the mass media that the oil company PDVSA is an environmentally clean company "that cares for the environment." Much before President Chávez, in Venezuela, the discussion on the negative impacts that the oil industry produces, such as air pollution, and water contamination by spills, has been considered taboo; currently, its publication through the mass media has been criminalized. Such was the case of the February 2012 oil spill in the Rio Guarapiche in Monagas State where the river and water sources of the population were severely contaminated.[25]

Socio-environmental objectives were left ambiguous in the national development plans, hindering their achievement; for example, the FSPSED did not establish environmental regulations regarding the mining sector with the exception of making some references to the use of natural resources under the principle of sustainability.

Under this Socialism of the 21st Century, the state took a more prominent role in the economy with the re-nationalization of many of the industries that were privatized in the 1980s. Currently, the Venezuelan government is not only the owner of state enterprises linked to the production of oil and other hydrocarbons, but is also playing a more decisive role than previous governments in creating the distributors' mechanisms of oil-rent to make its social, political, domestic, and international policies viable. The increasing socialization of the means of production by the state has given it more control, weakening instead of reinforcing the self-application of environmental regulations and controls.

Despite the greater emphasis the FSPSED gave to democracy and its relation to development and to the need for participation and organization, this plan does not include specific policies or criteria for the participation of affected communities in the framework of participatory democracy. The expected network of self-sufficient and democratic communities that at the local level have the objective of increasing environmental awareness and promoting a better environmental quality of life—called Communal Councils (Consejos Comunales) and Communes (Comunas)—have focused either on obtaining the government's financing to get infrastructural basic services at the local level or have been politically mobilized as Social Battle Units (Unidades de Batalla Social) in favor of President Chávez's third reelection. In any case, they do not have the required autonomy to build sustainable autonomous communities since they tend to respond to neopopulist and neoclientelist practices that stimulate political cooptation instead of having a voice in decision making.[26]

In the case of the regional agreements and megaprojects such as the IIRSA, they minimize environmental and sociocultural dimensions and there is also a sort of regional neopopulism that could lead to neocolonialism, since Venezuela makes claims to hegemony in the Caribbean region through distributing oil at subsidized prices. In addition, there is a contradiction between the proposed geopolitical integration to achieve sovereignty against anti-imperialism and the loss of territory and biodiversity, the forced displacements of communities including the indigenous nations, and the lack of respect for the constitutional environmental right that communities have to decide on their habitats.

Another unreal assumption is that development plans will stimulate endogenous development since most of the projects are linked to exogenous development. In fact, the current projects of physical interconnection, rather than integrating South American areas, are thought of as transport axes to reach ports on the Pacific Ocean and the Atlantic, or as Gudynas (2007) states:

> Roads are necessary to bring supplies to remote regions first, and then to be able to extract minerals, crops and other products . . . most of the exports are directed to destinations in other continents. Therefore, the so-called

> "physical integration" rather than joint production, promotes "interconnections" roads in order to reach the ports from where the ships to North America, Europe or Asia will depart. In the case of energy there are interconnections, such as pipelines, oil pipelines and overhead high-voltage rather than a common energy strategy.

Despite their magnitude and impacts, such projects have not been consulted with those affected, violating participatory democracy and the constitutional right to participation; this situation prompts the question of whether integration and regional sovereignty can be achieved at the expense of the environment and the constitutional rights that communities have to decide on their habitats.

The Social Movement's Resistance against Neo-extractivism

Since the beginning of this century, the Venezuelan socio-environmental movement mobilized around participatory democracy and the environmental, indigenous, and human rights that were included in the Constitution of 1999. The government's definition of the development model as "socialist" and its continuation of the extractive hydrocarbon economy led them to mobilize against this productive model. The growing social and political polarization of this century have weakened, demobilized, and divided the socio-environmental movement. The debate on the seizure of power as a prerequisite for the construction of a project of social transformation—which is very old within the Left—has divided the Venezuelan socio-environmental movement. While its defenders argue that political power is useful and necessary in any project of social transformation, its detractors warn that it concentrates forces opposed to any project of emancipation. Within this scenario, an important part of the former socio-environmental movement has been demobilized under the pretext that the institutionalization of environmental rights, together with the existence of a socialist government, are the guarantors of environmental and indigenous rights; therefore, the priority is to fight for the socialist revolution.

A second group corresponds to the scientific societies and the environmental nonprofit and nongovernmental organizations such as foundations. In the past, they also had the ability occasionally to behave like social movements when they mobilized and networked with the broader social movement around specific environmental demands;[27] nonetheless, given their need to get financing in a politically polarized setting, they currently concentrate their activities on carrying out censuses or environmental inventories or making the environmental impact studies for the government or for the private corporations. For this reason, this group is institutionalized in a way more permanent than before, identifying themselves with the sustainable development and green economy discourses.

A third group of activists have demobilized as they have joined the government's bureaucracy, accepting without question the guidelines and policies issued by the government, which prevents them from responding to the socio-environmental conflicts and demands; a few of them have joined the political arena as representative members in the national assembly. This group believes that political power is necessary to make the transformations leading to the socialism of the 21st century, even if there is the need to sacrifice autonomy. Additionally, there is a fourth group in charge of implementing the government's environmental projects, such as planting trees, environmental education, recycling trash, installation of energy saver lights, and so on. This group is becoming more institutionalized and it is considered the privileged interlocutor and the face of the government's policies, receiving the greatest coverage in the official media.

A fifth group, composed of some intellectuals and university professors, is critical of the negative effects of the government's policies despite its support of President Chávez; this group was the proponent of the 21st-century ecosocialism,[28] a term little accepted within the government because it is considered a petit bourgeois deformation, since for the government, the priority must be social redistribution and equity, not the environment. Some members of the group are the government's advisers and even worked in greening economic corporations such as the PDVSA.

The sixth group, who for the purposes of this chapter will be properly named as the socio-environmental movement, uses the strategy of mobilization to resist the productivist rationality involved in the current development model. This group is very heterogeneous in ideological terms since it is composed of both followers and opponents of President Chávez; on some occasions, the fifth group allies with this one, particularly in resisting extractivism. They give priority to the environmental and indigenous struggles over the ideological-political conflicts and assume that the extractive, predatory model of the environment is the same in industrial capitalism as in the socialism of the 21st century. This socio-environmental movement is the result of the coming together of at least three social movements: the indigenous, the environmentalist, and the human rights. However, not all indigenous or human rights groups are represented, as political polarization begins to permeate the indigenous and human rights movements. They propose a change of the paradigm and the rationality behind the government's developmental model and they mobilize demanding changes that transcend the capitalism-socialism government dichotomy. The fact that the government has criminalized protest and resistance from those who are against its policies has led this group to be the true actors of the environmental conflict in Venezuela.

By strengthening the extractive model, the Venezuelan government has many times forgotten the negative impact that it has on indigenous people, women, the environment, and the population in general. Like in other New Left countries of South America such as Ecuador or Bolivia, where social movements have challenged the extractive development model, the Venezuelan socio-environmental movement is resisting it from different positions, ranging from the defense of a healthy environment that involves a conception of sustainable development still based on anthropocentrism and the recently adopted slogan of the green economy as in the case of the multinational NGOs, to the visions of the *buen vivir* that involve a new rationality based on biocentrism and a democratic and autonomous way to achieve it: that is, the visions of the "other development," the "other democracy," and the right to "ancestral territoriality" sustained by the socio-environmental movements. In addition to mobilizing against the negative impacts of the extractive activities, another reason for the indigenous resistance is that they consider the government as violating the constitutional mandate to demarcate their lands and give them the property over their ancestral territories; and for them, such territorial demarcation is a prerequisite for the *buen vivir*. Geopolitical reasons, such as the safety of the border with Colombia or the existence of extractive mining and livestock settlements, are stated by the government as reasons for nondemarcation and dispossession from their ancestral lands of the indigenous communities of the Sierra de Perijá.

New Discourses, Same Developmental Paradigm

In Venezuela, the environmental, indigenous and human rights movement has been struggling to confront the intensification of the extractive model and the contradictions and conflicts derived from the inconsistencies between the constitutional environmental legal framework and the developmental emphasis of national plans on policies and projects. According to the socio-environmental movement the development paradigm did not changed with the arrival of President Chávez: "it's a change in the discourse but with the same old practices," the environmental organization AMIGRANSA declared, while the women and the environment organization GEMA stated that "the model continues to be focused on the exploitation of natural resources, oil, gold, coal . . . it focuses on the productivity instead of sustainability"; and the environmental organization ACOANA stated that "mining development continues to be the major strategy of this government . . . it still favors a model of development that is destructive to the environment."[29]

Arconada, a social movement activist but also a pro-Chávez supporter and an ex-governmental officer, criticized the continuation of the government's productivist approach in an open letter to President Chávez :

> CORPOZULIA and PDVSA (two of the largest stated owned enterprises) . . . pray without any shame the mining-extractives litany that if there is oil, it must be sought, if there is coal it must be gotten, if there is coltan, it must be extracted, if there is uranium, it must be sought, if there is gold, it must be mined . . . CORPOZULIA and PDVSA prompt an even more rentier economy, despite the environmental devastation that entails.[30]

In the counterforum of the World Social Forum that took place in Caracas in 2006, the Venezuelan socio-environmental movement drastically criticized the development model promoted by the government of President Chávez.[31] Protests and mobilization by pro- and anti-Chávez social movements were carried out during this forum in support of the protection of the territories and the habitat of the indigenous Wayuu, Yukpa, and Bari communities and against "the exploitation of coal" and "the neoliberal policy of the government despite its strong antineoliberal discourse."[32]

Sovereignty, Integration, or Neocolonialism?

One of the contradictions of the role assigned to oil and hydrocarbons within the socialism of the 21st century is that the extractive economy, which under the neoliberalism of previous governments served for the reproduction of relations of domination and exploitation, is now considered the basis for strengthening national sovereignty and the state.[33] For the socio-environmental movement, those goals are rhetoric. Homo et Natura, Soberania.com, and AMIGRANSA, among other social movements, accuse the Venezuelan government of not telling the truth about the energy megaprojects since the coal and gas expansion do not respond to endogenous-sustainable development as IIRSA representatives claim but to exogenous factors such as exportation as part of "the strategy of globalization of the World Bank (WB), the Inter-American Development Bank (IADB) and the Andean Development Corporation (CAF)."[34]

Another criticism by social movements is that the integration of products and markets proposed in IIRSA reinforces the subordinated role that South America has, and historically had, as supplier of raw material for large transnational industries. In relation to IIRSA and its goals of integration, Portillo remarks that, contrary to the goals of endogenous local sustainable development and integration, IIRSA's Western axis of development forms part of the strategy of the American Government and transnational corporations that seek to impose the Free Trade Area for the Americas (FTAA), the Panama Puebla Plan (PPP), and the IIRSA plan, in order to control South America's supply of natural resources in the world market; thus, the development of the Western axis, rather than the strengthening of the multipolar integration as desired, will benefit Colombia and Brazil and, through the latter, the United States, all of which will benefit more than Venezuela; on the other hand, the Eastern axis will benefit the Brazilian multinationals and their ally, the United States, more than Venezuela. Similarly, according to the environmental organization, AMIGRANSA, the requirements for an "endogenous-sustainable development" and for sovereignty are not met, since: "The IIRSA megaproject undermines the integrity of natural resources, jeopardizes the sovereignty of Venezuelan peoples and frontiers, and of other South American countries."[35] Instead of sovereignty and integration, the socio-environmental movement believes that "The Latin American Union (UNASUR) promoted by the Venezuelan Bolivarian process aims to integration but in practice ends up being neocolonialist when it uses Venezuela's energy and economic potential to become a new Latin American investor and lender country."[36] Moreover, for some environmental organizations, Venezuela's promotion of PETROCARIBE and PETROSUR is neocolonialist, because it promotes integration while using its energy potential to become a hegemonic country in Latin America.

The socio-environmental movement is resisting the exploitation of coal in the Sierra de Perijá under the argument that the negative socio-environmental impacts arising from its supporting infrastructure, instead of being "ecological mining activities," as the government says, will generate the deterioration of forests and produce scarcity of drinking water sources; additionally, they will sacrifice important economic activities of the area such as agriculture and livestock, thereby reducing food sovereignty, which paradoxically is another objective of the government. For their part, indigenous communities have highlighted their suffering from the socio-environmental impacts and the violation of collective human rights recognized by the Constitution. The Yutpas of the Sierra of Perijá, one of the indigenous communities most affected by the exploitation of coal in the Zulia State, pointed out that their most important resources for life are the resources of nature, water, plants, and animals, and that coal mining precludes the possibility of continuing with their traditional way of life and economic activities such as agriculture, breeding of animals, fishing, and hunting. It also reduces the availability of clean water from the rivers, threatens to displace them from their lands and makes the long-awaited demarcation of their ancestral lands increasingly difficult to achieve.[37]

Autonomy versus Cooptation

Despite the fact that by law the Venezuelan state is the owner of the natural resources and that the increasing expropriation of means of production reinforces its traditional control over the economy, the government has not self-regulated, generating conflicts with the affected communities and with the socio-environmental movement; moreover, contrary to the constitutional right to information, the government has attempted to hide critical information related to the negative environmental and sociocultural impacts. Within this scenario, the government has also tried to co-opt the socio-environmental movement and to appropriate discursively its struggles' symbolic flags. In some cases, the government has displayed populist policies as a substitute to constitutional participatory democracy, in an attempt to co-opt those social movements that oppose the extractive activities. This is particularly notorious in the indigenous communities that have been divided between those who accept "government's charity" and those who keep fighting for demarcation of their ancestral territories and against coal mining. As Portillo states:

> CORPOZULIA is working hard in the communities, giving away bags of food, killing 4 or 5 cows and inviting the people to eat, giving beers [. . . etc.]. All the missions in the region are under the tutelage of CORPOZULIA.[38]

When cooptation has failed, the government has tried to criminalize the protest and to neutralize it by excluding those social movements that do not endorse the socialist model of development or, like in the case of the indigenous population, that criticize its environmental, social or cultural impacts. It has contributed to dividing the previously integrated and autonomous socio-environmental movement on the basis of those who support the government's extractive development model, plans, and projects and those who oppose them. This criminalization has contributed to demobilizing an important and critical part of Venezuela's socio-environmental movement; moreover, the appropriation by the government of their symbolic demands has reduced their effectiveness and emptied their struggles of content. As a result, the previous Venezuelan socio-environmental movement is currently divided between the group that responds unconditionally to the government guidelines—and it includes the movement created as part of ALBA, which has been co-opted by the government in its attempt to conform to a "regional, revolutionary," and uncritical social movement—and the historical social movement that is resisting the extractivism on the basis that it violates their constitutional rights and their environmental and ancestral demands and struggles.

In addition, the existing political-ideological polarization has debilitated the socio-environmental movement, reducing its possibilities to use the media as a tool for resistance and as the way to transform its demands into "political facts" as it did in the past.[39] Neither the official nor the political opposition mass media, both deeply polarized, have any interest in disseminating the conflicts and issues posed by the socio-environmental movement. This has happened through alternative media, such as community radio networks or national or international electronic networks, where the socio-environmental movement has been able to politicize its struggles.

Participatory Democracy and Rio+20: Top-Down or Bottom-Up Decision Making

Participatory democracy has been predominantly discursive and restricted to the micro and, sometimes, local levels; President Chávez tended to make most of the relevant decisions at the national, regional, and international levels and had the last word concerning major policies and projects. Referring to this absence of participatory democracy in Venezuela, Arconada states: "Has the Ministry of the Popular Power questioned the capitalist essence of its hierarchical structure, and the privileges it entails?, does it build socialism from personal decisions which are not consulted with anyone, which do not follow any collective and democratic decision making process?"[40]

At the end of May 2012, the axis of social ecology of the Rosa Luxemburg Foundation in Venezuela convened interested persons and organizations and social movements for a two-day "Discussion on Rio+20: Sustainable Development and Green Economy." One of the objectives was to publicize the report of the International Coordination Group of the People's Summit (Grupo Internacional de Articulación de la Cumbre de Los Pueblos) and discuss the issue of the green economy in the light of the Venezuelan development model. Despite the fact that it was not the focus of the agenda, severe criticism of the extractive policies of the government was present in the discussions and debates and occasionally, it seemed to displace the central topic of the green economy. This event was not attended by any government official; indeed, less than a month before the summit of Rio+20, it was unknown what the official or government position would be, since the government was going to meet behind closed doors with advisers and official staff the day after this event.

In contrast to the second government of Carlos Andrés Pérez in 1992 and the first one of Hugo Chávez in 2002, the Rio+20 summit was the first of the three where the socio-environmental movement was not invited by the government to participate in the forum or in the official platform. Instead, the government was accompanied this time by advisers and government officials, some of whom in the past formed part of the socio-environmental movement and were critical of extractivism but today defend the idea that oil and hydrocarbons are a prerequisite for the socialist revolution of the 21st century.

While the discourse of the Venezuelan government strongly attacked the green economy, it signed the official document in the Rio+20 summit. On the other hand, the main strategy of the Cumbre de los Pueblos (People's Summit) to mobilize against the green economy impeded the alternative summit from focusing on the necessary debate about the negative socio-environmental consequences of hydrocarbons and the extractivist model of development. The low capacity of the Venezuelan socio-environmental movement to politicize these topics through the public and private media contributes toward a greater reduction in its importance in the national and government agenda.

National and Regional Resistance to Extractivism

Resistance to the hydrocarbons economy and to IIRSA transcends the national level; for instance, the IIRSA regional development plans and projects have been resisted by the autonomous environmental and indigenous movements in all South American countries, including Venezuela, since their inception in the year 2000. The Permanent Forum on Indigenous Issues that took place in the United Nations in 2008 and in which more than 300 indigenous people participated, urged governments, the World Bank and other multilateral and bilateral financial institutions to consider alternative systems of renewable resources as well as the low utilization of coal to replace the use of petroleum-based supplies.[41] Also, the World Peoples Conference on Climate Change and the Rights of Mother Earth celebrated in Cochabamba, Bolivia, in 2010 condemned the capitalist model because it "imposes mega infrastructure projects (sic) invades territories with extractive projects, privatizes water and militarizes the territory expelling the indigenous peoples and peasants from their territories, preventing food sovereignty and deepening the socio-environmental crisis."

The report of the International Coordination Group of the People's Summit for Social and Environmental Justice of 2012 "for unity and mobilization of peoples in defense of life and the Commons, environmental, and social justice against the commodification of nature and the green economy," was the product of a difficult consensus reached among social movements on the occasion of the Summit of Rio+20. This report is characterized by the lack of criticism of the extractive policies of the New Left governments. The document refers primarily to the mobilization strategy agreed against the green economy and the WTO, FTAA, and the G-20; paradoxically, the ALBA, UNASUR, and the G-77 were not mentioned.

The parallel and pro-official social movements created as part of the ALBA[42] and convened in its VI Summit of 2008 celebrated in Venezuela did not question the environmental and/or socio-cultural negative consequences of the mega-projects. This is because in Venezuela, the government not only has criminalized this sort of protest, but has left aside the autonomous social movements that question its policies and has stimulated the creation of an uncritical, parallel social movement to participate in ALBA, UNASUR, and the latest new regional institutions created. The leftist government's anti-imperialist and sovereignty discourses, which are similar to those of social movements, make it even more difficult for the social-environmental movement to move forward its agenda.

The lack of consistency between discourse and praxis seems to be present within the socio-environmental movement of the New Left countries. While the Latin American social movement has rejected discursively the extractive model in the latest social summits and the Cumbres de Los Pueblos and has proclaimed solidarity with those affected by the displacements and the negative impacts derived from the extractive projects and the supporting mega-infrastructures, the movement is divided with regard to its support of the New Left government's policies versus the indigenous struggles against the extractive activities. This fragmentation can be observed not only in Venezuela, where the government has set up its own parallel and ideologically allied social movement for participating in the ALBA and in other government institutions and social summits, but also in Bolivia and Ecuador where conflicts related with the extractive economy have intensified. New Left governments tend to identify those struggles as struggles against the socialism of the 21st century despite the fact that an important part of the mobilized communities tend to support the government.

Dilemmas and Challenges of Neo-extractivism and Socialism in the 21st Century

The socio-environmental movement not only questions the sustainability of the national and regional development plans but also their capability to be a valid alternative to capitalism as they are unable to incorporate their conceptions and new paradigms about nature and the "good living." The gap between discourses, plans, policies, and projects highlights the clear separation between the rhetoric of power and the dynamic process of the socio-environmental movement in Venezuela. Similar to previous governments accused of being capitalist and neoliberal, President Chávez, whose discourse claims his plans and policies are "revolutionary and anti-capitalist," has frequently adhered to a developmental vision that is not grounded in an objective evaluation concerning the negative impacts of the extractive model. Even though the government's social policies differ from neoliberalism because they emphasize the achievement of equity and more equitable social distribution and they are aimed to focus on the most vulnerable and unprotected population, its frame of reference is the neo-rentierism through the social missions and its purpose is to achieve social and political legitimacy, not the environmental sustainability of the supporting economic policies. Venezuela's 21st-century socialism shares with 20th-century socialism and with capitalism the notion of growth and progress and the endless development of the productive forces; in both cases, nature is subordinated to development.

Hydrocarbons in Venezuela, a discursive revolutionary-transforming tool to concrete the socialism of the 21st century, are supposed to contribute to sustainable-endogenous national development, the reinforcement of nationalism, regional integration and hegemony, independence from imperialism, the welfare and empowerment of poor people, and social and equitable inclusion among other objectives. Conversely, the fact that development plans, policies, and projects were designed top-down instead of bottom-up, which means they were not the result of constitutional participatory democracy, that they do not represent a collective construction of people's visions of a good life, and that they transgress basic environmental, indigenous, social, and human rights impeded the achievement of such aims.

The dilemma for New Left governments of the South America region is how to eliminate the dependency on the production and export of fossil fuels, which are considered as inherent to modern industrial capitalism, in order to achieve sustainable development and contribute to reduce global warming and other problems arising from the ecological crisis; and how to get enough income to maintain the domestic distributive social policies that have been very successful in achieving political and social legitimacy without accentuating the productive model. In the case of Venezuela, the dilemma is how to continue subsidizing oil prices to Caribbean and other Latin American countries in order to keep regional claims for hegemony without increasing the exploitation of fossil fuels and diversifying the economy when it sacrifices the environment and the right to the cultural identity of indigenous communities. Can we speak of justice, social equity, and solidarity or of a revolutionary-socialist democratic model that does not take into account the rights of the minorities who are usually the most vulnerable populations and does not stimulate environmental practices and relations that are more equitable with their communities and future generations? When it does not recognize the negative impacts of mega-projects whose costs are hidden for the sake of a new vision on Latin American integration around the ALBA, UNASUR, or other regional institutions? Is it possible to build a new socialism that overcomes the mistakes of the past, incorporates environmental and social sustainable alternatives to development, stimulates participatory democracy and includes all actors? An open final question is how to "overcome development" and neo-extractivism as some scholars have suggested and move toward an alternative to development, post-development, or "decolonial" projects.[43]

The answers to these questions should be built collectively at the national and regional levels, through the exercise of participatory democracy mechanisms such as referendums where everyone can participate. On the other hand, the Venezuelan Ecosocialism Forum of 2007 warned about the need for a critical analysis of 20th-century socialism's experiences to regain the credibility of socialism as an alternative project to capitalism, and it recommended changing the cultural paradigm since the rationality or developmental logic of the productive model has not changed. Neither the ecosocialist discourse nor the development plans have been able to transcend their discursive level in order to manage the arising environmental contradictions and conflicts to be able to incorporate a biocentric vision of nature. Despite the existence of a constitution that legitimizes environmental and indigenous rights, an anticapitalist alternative named the socialism of the 21st century, and development plans that incorporate a diversity of articles in favor of sustainability, in Venezuela—similar to other New Left countries in South America—environmental requirements are perceived (by those governments) as Gudynas has suggested,[44] as obstacles to economic growth and they are considered a brake for the reproduction of the state apparatus and the economic assistance to the neediest. For all these reasons, the productivist, instrumental, and developmental logic of the extractive exporting development model, which attempts to break with the past colonialist legacy, has been unsuccessful in facing the increasing contradictions and conflicts.

Mobilizations against neoliberalism and the anti-imperialism in Venezuela have not played the transforming and critical role assigned to social movements. While President Chávez won the first election in 1998 with the votes of a diverse ideological and social base that included mostly Left-wing groups and above all the popular sectors, some of the Left-wing groups, including many of the intellectuals, distanced themselves. The fact that the presidency of the republic has tried to organize the "people or the sovereign" could jeopardize the autonomy of social movements, including the environmental one. Another question is to what extent such organizations can respond to the anti-imperialist and anti-neoliberal mobilizing logic that presupposes the existence of an autonomous social movement that goes beyond the local and micro-local levels of those organizations that have a high potential to be co-opted by the state through financing. While it is true that there are social organizations promoted by the government that can respond to the logic of social movement such as the Movimiento Ezequiel Zamora, the Asociación Nacional de Medios Comunitarios, Libres y Alternativos (AMCLA), and the urban land committees among others, most of them, such as the community councils or the ones linked to social missions, can hardly conform to the notion of autonomy required by social movements.

Currently in Venezuela there is not a social movement with a transformative potential. It is plausible that given the ideological heterogeneity of supporters of Chávez and given the lack of a shared ideological project within the Chávez movement, strategies that do not necessarily have neoliberalism as a reference point might be displayed. This could be the case of the socio-environmental movements because while they are anti-neoliberal by definition and many of their members support Chávez, they transcend this dichotomy to the extent that they question the civilization model and demand the transformation of the political, cultural, gender, and environmental social and rationality. The real revolution is not only at the economic and social levels but also at the cultural, environmental, and gender levels; this means there is a need for change in the civilization model. In this regard, the Bolivarian model can generate resistance and mobilization on the part of the anti-neoliberalism and anti-neoglobalization social movements on the grounds of being an environmentally predatory model that causes high socio-environmental negative impacts.

Acknowledgment

The author wants to thank Adela Alfonsi for her useful and amicable editorial help which contributed to making the text readable.

Notes

1. Author's free translation. Eduardo Gudynas "La izquierda marrón." (2012). http://www.cetri.be/spip.php?article2552lang=es (Accessed April 22, 2012).

2. See García-Guadilla, "Environmental Movements, Politics, and Agenda 21." As in the 1992 Rio Summit, in the debate of the official Latin American and Caribbean Platform for Johannesburg 2002, which was part of the regional forum of governments, NGOs and social movements strongly rejected the economic model behind sustainable development because they considered it a new attempt to save the old, unsustainable neoliberal and capitalistic model of economic development. In contrast, the G-77 governments, at that time coordinated by Venezuela, made efforts to skip the hydrocarbons discussion as it has done in most of the environmental summits.

3. Acosta, "El Buen Vivir," and Gudynas, "Buen Vivir: Germinando alternativas de desarrollo."

4. Banco Central de Venezuela, "Exportaciones e importaciones."

5. Perroux, "A Note on the Notion of the Growth Pole."

6. Friedman, *Regional Development Policy;* and Rodwin, *Planning Urban Growth.*

7. García-Guadilla, "The Old Story of New Cities," and "Impactos socio-económicos, políticos y espaciales."

8. Rodríguez, C. "Conflictos indígenas," and García-Guadilla, "El movimiento ambientalista."

9. Méndez, "Conflicto sociopolítico y socio-ambiental," and Buroz, "Concepciones y racionalidades."

10. Rodríguez, L. "El conflicto socioambiental."

11. República Bolivariana de Venezuela, "Plan de Desarrollo Económico y Social de la Nación 2001–2007" (SENPF); and "Proyecto Nacional Simón Bolívar: Primer Plan Socialista" (FSPSED). Since the FSPSED is an abridged plan, it presupposes the more extensive guidelines of the SENPD.

12. See http://www.pdvsa.com/index.php?tpl=interface.sp/design/readmenuprinc.tpl.htmlnewsid_temas=44

13. President Chávez called for the creation of social programs in 2003 that came to be known as *Misiones*—"missions"—and were crafted to be displayed in Venezuela's poorest areas and encourage participation from members of the communities in which they were present. What started as a small number of missions in 2003 has grown to close to 26 missions today covering a large number of social needs. See http://venezuela-us.org/economia-y-desarrollo-social

14. PDVSA, "A National Popular and Revolutionary Oil Policy."

15. Pérez, "La Revolución Bolivariana. La vía venezolana hacia el desarrollo sustentable."

16. See http://www.pdvsa.com/index.php?tpl=interface.sp/design/readmenuprinc2.tpl.htmlnewsid_temas=38)

17. The *consejos comunales*, "communal councils," are social organizations created in 2006 by presidential initiative to carry out the necessary changes for the Bolivarian revolution and for the socialism of the 21st century. The main activity of the 40,629 communal councils (see Caione, "Creación de las comunas," 18) has been the formulation of projects that provide benefits to its territorial habitat in order to receive the resources offered by the government for the improvement of the quality of life and community services. See García-Guadilla, "La praxis de los consejos comunales," and Alvárez and García-Guadilla, "Contraloría Social y clientelismo."

18. See http://www.cepal.org/usi/noticias/bolfall/7/38207/FAL-280-WEB.pdf

19. See http://www.cepal.org/usi/noticias/bolfall/7/38207/FAL-280-WEB.pdf

20. IIRSA's strategic role consists of overcoming the obstacles of physical integration to foment intraregional commerce in South America, to stimulate the reorganization of productive chains, to contribute to build an integrated, competitive and dynamic South American economy in a framework of social and environmental sustainability, to stimulate the participation of the private sector, to harmonize public policies and regulatory framework between the countries and sectors, and to diminish the distribution and commercial cost through the development of infrastructure in transport, energy, and telecommunications sectors (See http://www.cepal.org/usi/noticias/bolfall/7/38207/FAL-280-WEB.pdf)

21. IIRSA, "Cartera de Proyectos," "Base de datos de proyectos," and "IIRSA 10 años después."

22. PDSVA, "Informe de Gestión Anual," 2009, 2010, and 2011; and "A National Popular and Revolutionary Oil Policy."

23. COSIPLAN, "Plan de Acción Estratégico," and "Consejo Suramericano de Infraestructura y Planeamiento."

24. República Bolivariana de Venezuela, "Plan de Desarrollo Económico y Social de la Nación 2001–2007," p. 82–83.

25. See http://informe21.com/actualidad/12/02/10/emergencia-en-monagas-por-derrame-de-petroleo-en-el-rio-guarapiche-fotos, http://www.soberania.org/Articulos/articulo_7060.htm, http://amigransa.blogia.com/2012/febrero.php, http://periodicoellibertario.blogspot.com/2012/02/ecocidio-petrolero-en-monagas-revela.html

26. Arenas and Gómez Calcaño, *Populismo autoritario*; Roberts, *Populism, Political Conflict, and Grass-Roots Organizations in Latin America*; García-Guadilla, "La praxis de los Consejos Comunales"; and Alvarez and García-Guadilla, "Contraloría Social y clientelismo."

27. García-Guadilla, "The Venezuelan Ecology Movement."

28. Fergusson, "El desarrollo endógeno sustentable"; Fergusson, "Venezuela: *la cuestión ambiental*"; and Velasco, "Participación en el Foro Ecosocialismo," 2005.

29. GAUS, "Base de datos de entrevistas."

30. Arconada, "Carta abierta al Presidente Hugo Chávez."

31. García-Guadilla and Lagorio, "La cuestión del poder."

32. For conflicts between the socio-environmental movement and the government on these topics for the period 2004–2012, see webpages: htpp://www.soberanía.org.ve, http://www.soberania.org/lusbi_portillo_portada.htm, and http://www.rebellion.org.ve, http://elistas.egrupos.net/lista/lea/archivo/indice/519/ msg/929/

33. Lander, "El estado de los actuales procesos."

34. Some of the criticism from the socio-environmental movement (Homo et Natura, Federación Ecologista del Zulia, Frente Nacional por la Defensa del Agua y la Vida) against the hydrocarbons projects such as coal mining and IIRSA can be seen in: www.Soberania.org/lusbi_portillo_portada-htm. The criticisms from human rights organizations such as PROVEA (2004–2012) and La Red de Apoyo por la Justicia y la Paz for the period 2004–2012 can be seen in: www.derechos,org.ve and www.redapoyo.org.ve

35. GAUS, "Entrevistas a organizaciones ambientalistas."

36. See http://www.soberania.org/lusbi_portillo_portada.htm and http://www. rebellion.org.ve

37. Rodríguez, "El conflicto socioambiental." All over Latin America, researchers have alerted about the unsustainability of IIRSA, pointing out that the intensive extraction of nonrenewable natural resources, which causes the destruction of biodiversity-rich areas and massive deforestation, results in irreversible damage to ecosystems, damage to indigenous territories, and the displacement of thousands of people. It also contributes to loss of biodiversity, loss of livelihood, worsening of poverty and in short, risking the survival of future generations and affecting the last existing natural reserves that exist not only in Latin America but also the planet.

38. http://www. Soberania.org\articulo_3557.htm

39. García-Guadilla, "The Venezuelan Ecology Movement."

40. See, http://www.aporrea.org/ideologia/a116888.html

41. Foro Permanente para las Cuestiones Indígenas en las Naciones Unidas. See: http://alainet.org/active/31235lang=es and http://www.cinu.org.mx/prensa/especiales/indigenas/noticias.htm

42. Bossi, "Los movimientos sociales en el Alba."

43. Escobar, "Latin America at a Crossroad."

44. Gudynas, "Estado Compensador y Nuevos Extractivismos."

Bibliography

Acosta, Alberto. "El Buen Vivir, una utopía por (re)construir." *Revista Casa de Las Américas* 257 (2009): 33–46.

Alvárez, Rosangel, and María Pilar García-Guadilla. "Contraloría Social y clientelismo: la praxis de los Consejos Comunales en Venezuela." *Politeia* 34, no. 46 (2011): 175–207.

AMIGRANSA. "Insostenibilidad ambiental del modelo de desarrollo Bolivariano: Programa." *Foro Alternativo al Social Venezolano* blog, January 27, 2006. http//www.soberanía.org

Arconada, Santiago. "Carta abierta al Presidente Hugo Chávez Frías: El otro diálogo." February 2, 2011. http://www.aporrea.org/ideologia/a116888.html (Accessed March 2, 2012).

Arenas, Nelly, and Luis Gómez Calcaño. *Populismo Autoritario: Venezuela 1999–2005.* Caracas: Universidad Central de Venezuela, 2006.

Banco Central de Venezuela. "Exportaciones e importaciones de bienes y servicios según sectores." *Balanza de pagos y otros indicadores del sector externo.* Caracas: Banco Central de Venezuela, 2011. http://www.bcv.org.ve/c2/indicadores.asp (Accessed May 27, 2012).

Bossi, Fernando. "Los movimientos sociales en el Alba." *Aporrea* blog, August 1, 2008. http://www.aporrea.org/venezuelaexterior/a61425.html (Accessed July 15, 2012).

Buroz, María Teresa. *Concepciones y racionalidades del desarrollo sostenible. Caso: la apertura minera en la reserva forestal de Imataca, Venezuela.* Diss., Caracas: Universidad Central de Venezuela, 2001.

Caione, Cecilia. "Creación de las comunas sufrió tropiezos." Diario *Ultimas Noticias* (Venezuela). June 20, 2011. http://www.ultimasnoticias.com.ve/noticias/actualidad/politica/creacion-de-comunas-sufrio-tropiezos.aspx

COSIPLAN. Consejo Suramericano de Infraestructura y Planeamiento. (2012) "Plan de Trabajo 2012 COSIPLAN-IIRSA." www.iirsa.org/ . . . /rc_brasilia11_6_plan_de_trabajo_2012.pdf

COSIPLAN. "Plan de Acción Estratégico 2012–2022." UNASUR, 2011.

Escobar, Arturo. "Latin American at a Crossroad. Alternative modernization, post-liberalism, or post-development?" *Cultural Studies* 24 (2010): 1–65.

Fergusson, Alex. "El Desarrollo Sustenable. La revolución de las lógicas y los sentidos." 2007. http://happytreeflash.com/endogeno-ppt.html (Accessed July 24, 2012).

Fergusson, Alex. *Venezuela: la cuestión ambiental y el desarrollo.* Caracas: Ediciones del MPPCyT, 2008.

Friedman, John. *Regional Development Policy: A Case Study of Venezuela.* Cam bridge: MIT Press, 1966.

García-Guadilla, Maria Pilar. "El movimiento ambientalista y la constitucionalización de nuevas racionalidades: dilemas y desafíos." *Revista Venezolana de Economía y Ciencias Sociales* 7, no. 1 (2001): 113–32.

García-Guadilla, María Pilar. "Environmental Movements, Politics and the Agenda 21 in Latin America." United Nations Research Institute for Social Development, 2005. http://www.unrisd.org/80256B3C005BCCF9/(httpPublications)/BC70FF0CD26D614AC12570AE003085A9?OpenDocumentlanguage=en

García-Guadilla, María Pilar. "Impactos socioeconómicos, políticos y espaciales de las grandes inversiones minero-industriales en América Latina: Aproximación teórico-metodológica." *Revista Interamericana de Planificación* (SIAP) XXI no. 81 (1987): 15–43.

García-Guadilla, María Pilar. "La praxis de los Consejos Comunales: Poder Popular o instancia clientelar?" *Revista Venezolana de Economía y Ciencias Sociales* 14, no. 1 (2008): 125–51.

García Guadilla, María del Pilar. "The Old Story of New Cities: Unanticipated Social Costs of Ciudad Guayana, a Planned Industrial City in Venezuela." PhD diss., University of Chicago, 1985.

García-Guadilla, María Pilar. "The Venezuelan Ecology Movement." In *The Making of Social Movements in Latin America,* edited by Arturo Escobar and Sonia Alvarez, 150–70. Boulder, CO: Westview Press, 1992.

García-Guadilla, María Pilar, and Carlos Lagorio. "La cuestión del poder y los movimientos sociales: Reflexión post-foro social mundial Caracas 2006." *Revista Venezolana de Economía y Ciencias Sociales* 12, no. 3 (2006): 53–65.

GAUS. "Base de datos de entrevistas a las organizaciones de derechos humanos, ambientalistas y de democracia liberal." In *La constitucionalización de nuevas ciudadanías y racionalidades en Venezuela: actores sociales y gestión de conflictos sociopolíticos,* edited by María Pilar García-Guadilla. Grupo Interdisciplinario de Investigación en Gestión Ambiental, Urbana y Sociopolítica, (GAUS), Universidad Simón Bolívar-FONACIT. Vol. 3, Part I, 1–258, Caracas, 2010.

Grupo de Articulación Internacional de la Cumbre de los Pueblos por la Justicia Social y Ambiental. (International Coordination Group of the People's Summit for Social and Environmental Justice). "Informe del Grupo de articulación internacional de la Cumbre de los Pueblos por la Justicia Social y Ambiental." May 12, 2012, Rio de Janeiro.

Gudynas, Eduardo. "Buen Vivir: Germinando alternativas al desarrollo." *América Latina en Movimiento,* ALAI, 462 (Feb 2011): 1–20.

Gudynas, Eduardo. "Estado Compensador y Nuevos Extractivismos: las ambivalencias del progresismo suramericano." *Nueva Sociedad* 37 (2012): 126–46.

Gudynas, Eduardo. "La Sopa de la integración." In *Integración Sur* blog. 2007. http://www.integracionsur.com/americalatina/GudynasSopaIntegracion.htm (Accessed May 29, 2012).

Harneker, Marta. "De los consejos comunales a las comunas. Construyendo el Socialismo del Siglo XXI." 2009. http://es.scribd.com/doc/27443484/Harnecker-Marta-De-los-consejos-comunales-a-las-comun-2009 (Accessed May 18, 2010).

IIRSA. "Cartera de Proyectos IIRSA." (2011). http://www.iirsa.org/Cartera.asp

IIRSA. *IIRSA 10 años después: Sus logros y desafíos.* 1ª Edición–Buenos Aires: BID-INTAL, 2011. http://www.iirsa.org/BancoConocimiento/I/iirsa_10_anos_despues_sus_logros_y_desafios/iirsa_10_anos_despues_sus_logros_y_desafios.asp?CodIdioma=ESP

IIRSA. "Listado de Proyectos ordenados por Eje y Código." (2011). http://www.iirsa.org/proyectos/Proyectos.aspx (Accessed June 23, 2012).

Lander, Edgardo. "El Estado en los actuales procesos de cambio en América Latina: Proyectos complementarios/divergentes en sociedades heterogéneas." In *Mas allá del Desarrollo,* edited by Miriam Lang and Dunia Mokrani, 121–43. Quito: Fundación Rosa Luxemburg/Abya Yala, 2011.

Latin American Commission on Development and Environment (LACCDE). *Our Own Agenda.* Washington: Interamerican Bank, 1990.

Méndez, Diana. "Conflicto sociopolítico y socio-ambiental de la apertura minera en la Reserva Forestal de Imataca." Hon. diss., Universidad Central de Venezuela, Caracas, 1999.

PDVSA. "Informe de Gestión Anual." PDVSA, Caracas, 2009. www.pdvsa.com/interfase.sp/database/fichro/free/5889/1049.PDF (Accessed May 27, 2012).

PDVSA. "Informe de Gestión Anual." PDVSA, Caracas, 2010. http://es.scribd.com/doc/51489238/PDVSA-Informe-Anual-Gestion-2010

PDVSA. "Informe de Gestión Anual." PDVSA, Caracas, 2011. http://static.eluniversal.com/2012/04/17/informedegestion.pdf (Accessed May 27, 2012).

PDVSA. "A National Popular and Revolutionary Oil Policy." *Ministerio de Energia y Petróleo.* PDVSA. Full Sovereignty over Oil. Series no. 1, 2006. http://www.pdvsa.com/interface.en/database/fichero/publicacion/1413/61.PDF

Pérez, Jesús A. "La Revolución Bolivariana. La vía venezolana hacia el Desarrollo Sustentable," 2007. http://www.vision-latina.ca/index.php?view=articlecatid=36%3Aeditorialesid=58%3Ala-revolucion-bolivariana-la-via-venezolana-hacia-el-desarrollo-sustentableformat=pdfoption=com_contentItemid=53 (Accessed May 1, 2009).

Perroux, Francois. "Notes sur the pole de croissance" A Note on the Notion of Growth Pole. *Economie Appliquee* 7, no. 1–2 (1955): 307–20.

Portillo, Lusbi. "Venezuela perderá su soberanía en los proyectos de la IIRSA." In *Aporrea* blog, October 17, 2004. http://www.aporrea.org/tiburon/n51606.html (Accessed June 25,2012).

Programa Venezolano de Educación-Acción en Derechos Humanos (PROVEA). "Informes Anuales 2004–2011." http://www.derechos.org.ve/ (Accessed May 27, 2012).

Red de Alerta Petrolera/Orinoco Oilwatch. "El Plan Estratégico de la Nueva PDVSA: ¡Un nuevo salto al abismo¡". In *Soberanía.org* blog, July 17, 2005, Caracas. http://www.soberania.org/Articulos/articulo_1474.htm (Accessed May 27, 2012).

República Bolivariana de Venezuela. "Proyecto Nacional Simón Bolívar, Primer Plan Socialista. Desarrollo Económico y Social de la Nación 2007–2013." http://www.bibliotecasvirtuales.org.ve/spaw/docs/Mapa_Lineas_General_Econ_Social_Plan_2007_2013.pdf (Accessed February 11, 2011).

República Bolivariana de Venezuela. *Constitución de la República Bolivariana de Venezuela.* 1999.

República Bolivariana de Venezuela. "Plan de Desarrollo Económico Social de la Nación 2001–2007." Caracas, Venezuela, 2001. http://www.mpd.gob.ve/pdeysn/plan.htm (Accessed February 11, 2011).

Roberts, Kenneth. "Populism, Political Conflict, and Grass-Roots Organization in Latin America." *Comparative Politics* 38, no. 2 (January 2006): 127–48.

Rodríguez, Claudia. "Conflictos indígenas y ambientales en el Estado Bolívar." MA diss., Universidad Simón Bolívar, Caracas, 2001.

Rodríguez, Laura. "El Conflicto socioambiental en la Sierra de Perijá: Concepciones sobre el desarrollo y alcances de la democracia ambiental (2000–2010)." MA diss., Universidad Simón Bolívar, Caracas, 2012.

Rodwin, Lloyd. *Planning Urban Growth and Regional Development. The Experience of the Guayana Program of Venezuela.* Boston: MIT Press, 1969.

Salcedo, Paola. "Hechos sobre el derrame petrolero en Jusepin." In *Amigransa* blog, February 18, 2012. http://amigransa.blogia.com/2012/022102-hechos-sobre-el-derrame-petrolero-en-jusepin-por-conflictove.php (Accessed June 26, 2012).

Uzcátegui, Rafael. "Foro Social Mundial Caracas: mortaja para los movimientos sociales." *El Libertario* blog, December 3, 2005. http://colectivonph.com.ar/nph/index2.php?option=com_contentdo_pdf=1id=220 Accessed May 27, 2012.

Velasco, Francisco Javier. "Ecotopia y Socialismo para el Siglo XXI." *A Plena Voz: Revista Cultural de Venezuela* 18 (2005): 16–19.

Velasco, Francisco Javier. "Participación en el Foro Ecosocialismo del Siglo XXI," PDVSA (lecture, recorded), Caracas, May 2007.

World Commission on Environment and Development. *Our Common Future.* Oxford: WDED, 1987.

Chapter 5

Environmental Justice Networks in Latin America: Compared Experiences in Argentina, Brazil, and Mexico[1]

Mauricio Berger

Introduction

In Latin America today, governments and the national and transnational corporations are enforcing the exploitation of the common goods and the destruction of forms of life and environment for economic development and the new requirements of capital accumulation. In the history of Latin America commodification has been a constant, but this model of production/destruction advanced remarkably during the decades of neoliberal globalization. The region's participation in the exploitation of minerals, water, and biodiversity was not only consolidated but deepened with technological innovations and a patent regime that allows private ownership of common goods such as native seeds and even human genetic information.

Very close to the key definition by David Harvey of accumulation by dispossession, this is a process of privatization of common goods such as water, forests, minerals, land, and knowledge. "The inability to accumulate through expanded reproduction on a sustained basis has been paralleled by a rise in attempts to accumulate by dispossession,"[2] the author explains, is the permanent plundering and robbing of people's rights.

Even though the governments in the region affirm the environmental sustainability of these policies, environmental inequities are broadened. While in the decade of the 1990s the social conflicts were related to the hegemonic neoliberal policies (effects of privatization of public services, deregulation of labor rights, and precariousness of living conditions, among others), in the present the environmental conflicts multiply around the devastating effects of large-scale mining, agrobusiness, and mega-infrastructure projects: the destruction of forms of life and their environments, massive diseases and deaths related to pollution, and degradation of soils and water, among other things.

Because of this systematic plundering, there are many instances of environmental injustice in which governments and their institutions have failed and/or refused to provide protection for the rights of the affected peoples. This context shapes a political practice of networking, basically understood as a form of self-organization that evolves from connections and solidarities between a multiplicity of individuals and collectives in their fight for rights and justice. The emergence of local struggles through the formation of networks can be understood then as the self-defense practices of affected communities. In doing so, the creativity of their resistance undermines[3] the pretension of governments and corporations to rule over life, health, and environment without the consent of the affected peoples. While the pretension of this capital sovereignty over the lives and the rights of the peoples is possible through the functioning of a network of corporations, public officials and institutions, media, universities, NGOs and other organizations that spread the control and exploitation of the common goods, the struggles have also taken the form of networking through a creative cooperation of diverse kinds of knowledge, values, practices, and forms of life. The ruling power cannot be imposed with impunity because of the resistance of the affected peoples, which is proof of the generation of a constituent power from below.

In this context, Environmental Justice Networks (EJN) can be understood as forms of organization that evolve primarily against environmental injustices due to extractivist corporations and the complicity of state power. Even though EJNs are spread all over Latin America, this article will be based on three experiences: the Brazilian Network of Environmental Justice, the National Assembly of Environmentally Affected People in Mexico, and the Network of the so-called Fumigated Peoples (Pueblos Fumigados) in Argentina.

The article then aims to contribute to the public understanding of EJN in Latin America. In part one, a brief description of their frameworks, organizational forms and actions in the Brazilian, Mexican, and Argentinean experiences is presented. In part two, there follows a work-in-progress comparative analysis in order to think through EJN's common aspects and political power.[4]

This approach to the present of the environmental issues in Latin America provides not only an explanation on the side of the structural generation of environmental injustices but mainly a comprehension of the political practices on the side of environmental justice (EJ). The several situations of unequal distribution of environmental risks and damages are not accepted without opposition. The claims for recognition and protection of rights to life and health over economic profit develop strategies of self-defense by means of an institutional creativity[5] that should be considered as keynotes from Latin America for the comprehension of global environmentalism and its challenges.

Part 1: EJN Experiences in Latin America

Brazil: The Experience of the Brazilian Network of Environmental Justice (Rede Brasileira de Justiça Ambiental—RBJA)

In Brazil, the crisis/reaccomodation of neoliberalism coexists with the transition to neo-developmentalism and neo-extractivism in the region. The Brazilian Network of Environmental Justice emerges then as a space of convergence of different critical visions and positions about the predatory model of development in the rise of Brazil as a new capitalist power nation. This model is particularly damaging to individuals more at risk because of their class, ethnicity, or gender, since the polluting industry complex and polluting activities are localized in working-class neighborhoods and black and indigenous communities all over the country.

Within this context, the RBJA had its initial moment by the end of 2001 when the International Colloquium on Environmental Justice, Labor and Citizenship was held involving organizations of civil society, social movements, academics, workers' unions, and members from the U.S. EJ Movement. As Julianna Malerba—RBJA former national secretary—recalls, the visit of the U.S. EJ organizations was considered a show of international solidarity to stop the exportation of environmental damages from U.S. companies to the global South.

The RBJA constitutes itself with the objective of providing public visibility and strengthening the movements, collective and individual struggles, against environmental injustice. The network builds a space for exchanging information, solidarities, and public accusations.[6] The articulation and mobilization strategy gathers almost 300 organizations coming from a diversity of political trajectories, identities, and organizational forms. Affected peoples' associations and movements, indigenous communities, Quilombolas (people who live in settlements called quilombos), academic and professional centers, environmental NGOs and networks, workers' unions, women's movements comprise among others, the diversity of the RBJA.

As Henri Acselrad—one of the founder members—points out, the RBJA is a result of a collective construction:

> RBJA arose from the need to organize a field of social struggle from the perspective that there is no dissociation between the social and environmental issues. If there's something remarkable since the Conference Rio 92, it is the constitution of that field as a result of the analysis made by movements and academics related to them (. . .) To become part of the network was an opportunity for the exercise of discussing strategies and exchanging information and experiences among those who expect to produce a collective and autonomous intelligence.[7]

With regard to the framework of environmental justice, Marcelo Firpo affirms that "the emergence of the EJ debate in Brazil contributed to link environmental problems with public health, human rights and justice."[8] Firpo points out some differences between the process that led to the development and the usage of the EJ concept in the United States and in Brazil. In his words, while the U.S. EJ movements were strongly organized by local communities and specific ethnic groups, in Latin America and Brazil, EJ discussions leveled criticism against the capitalist development model and the unfair regional role within the context of international trade since their very beginning. He states that beyond singular and historical patterns of social inequality and ethnic discrimination, environmental conflicts in Latin America have a strong relationship with their insertion of the region's countries in the international economy. Yet he adds: "Another significant aspect in Brazil is the relevance of traditional communities—indigenous, quilombolas (African slaves' descendants established in country areas), and other native people who live from forest gathering, agriculture, and fishing—inhabiting territories disputed by powerful economic groups from the industries of agribusiness, mining and construction of hydroelectric dams."[9]

As the RBJA's Declaration of Principles states, by the term "environmental justice" they understand the set of principles and practices that:

a. assures that no social group, be it ethnic, racial or class oriented should bear a disproportionate share of the negative environmental consequences of economic operations, policy decisions and federal, regional and local programs as well as the lack of such programs or the neglect of their enforcement;
b. guarantees fair and equal access, directly or indirectly, to the environmental resources of the country;
c. assures full access to relevant information about the use of environmental resources, the destination of wastes, the sites which are sources of environmental hazards, as well as assuring participatory and democratic procedures in the definition of policies, plans, programs and projects concerning such issues;
d. enables the empowerment of collective actors, social movements and popular organizations so that they become protagonists in the building-up of an alternative pattern of development which might assure the effective democratization of the access to environmental resources and their sustainable use.[10]

During its first 10 years, the RBJA has functioned with a national secretary who was in charge of the coordination of the network activities. There was no central command, because every member would speak on behalf of his organization and the collective declarations would be decided through discussions in the RBJA national encounters and also through the exchanges on the web forums.

Though the lack of a formal structure was sometimes considered as a limitation to the network's extension and political impact, it was still kept as a guarantee of plurality. There was some kind of pressure to define a formal representation as RBJA was constantly called to send its representatives to encounters and spaces with political incidence impact, such as public hearings, forums, and environmental agencies. Therefore, one permanent debate within the RBJA is about representation (representatividade), the position of the network as a political actor toward state institutions and environmental agencies, and the role of the network in collective decision making.

RBJA members were aware that a centralized structure for which decision making tends to be vertically concentrated would generate difficulties considering the diversity of its composition and only few issues would have an expression of the network position. One learning they produced was to combine the networking form with the autonomy of its nodes. The creation of working groups instead has permitted autonomous movements, with the RBJA not being able to make public declarations and actions on behalf of everyone because of the wide range of positions encompassed within it.

Through its experience, the network has also learned from dilemmas such as horizontality and diversity versus an effective decision making process. Which spaces are best for collective decision making? Which criteria for a consensus that reflects the most the diversity of opinions and positions? With some of these questions in permanent debate, after 10 years of existence the RBJA voted in its last encounter the transition to an organizational form called the political council (*colegiado político*). As Raquel Rigotto and other members of the RBJA say in the IV National Encounter Report, this form of organization is thought by the RBJA members as an experience within the process of construction of the network, the support for a horizontal group of grassroots movements, entities, and academia, and the potentiality of mutual empowerment. It recognizes the limits of the previous structure based on a national secretary to encourage a process of expansion of the collective potentiality, make decisions in a collective way, and promote articulations and deliberations with all the members.

> The Colegiado encourages the respect, valorization and cultivation of a diversity of ethnic, gender, ecosystem, struggles, political positions. It also aims to invest efforts in the creation of collective deliberative forms, democratic, decentralized, sharing responsibilities between all the members, a group that takes care of horizontality and its continuity.

The issues about networks' organizational form are not only the ones related to a structure of representation. One major contribution from the RBJA is the reflection about the spatial and temporal limits of networks. As Marcelo Calasanz, one of the RBJA members, suggests:

> Maybe we have to reconsider the creation of perennial and continuous processes and give place to non-lineariity and oscillation of collective action movements and their political moments according to the circumstances. There are temporal and spatial limits of networks, because networks add new territories and new issues and may get to a collapse of representation.[11]

In this regard, Calasanz asks: "To what extension does geographical dispersion and diversity affect the collapse of the network management?" He proposes that networks do not represent, they are not movements, they articulate diverse processes. There is a complex articulation that allows the specific local and regional scales to not be forgotten when they skip to the national and transnational scale. In so many cases there is an abrupt jump from local to global without any mediation between these transitions. The local cannot follow this jump and remains unarticulated, isolated, without constructing the necessary resonance at the grassroots level, so the global does not have a rearguard. Networks articulate, they do not replace local struggles, and one main concern here is to keep the protagonism in every scale, in how a local struggle can get involved in a global scale and how networks support and empower local struggles. Calasanz also warns about some obstacles that networks face, such as the temptation of becoming an NGO (managerialism, verticalization, need to meet established goals, results indicators), which facilitates the undermining of networks. In this sense, irregularity is a virtue, because if we were "lineal," continuous, perennial and constant, we would be easily mappable, identifiable, and deconstructed.

Given the multiplicity existing in the RBJA, another point here is that the network's consensus/deliberative spaces require an inclusive communication, with a language that anyone can have access to, not just the experts. This calls for a permanent practice of translations toward the use of common terms. Also, the reaffirmation of the transversality of race, gender, and class in environmental injustices even inside the network is considered essential in order to avoid exclusions of any kind. Not only an inclusive language and translations are needed, but also a platform of common agreements. In this regard, Henri Acselrad affirms that those who reach an understanding of the configuration of the political space in which the RBJA is disposed to act perceive the privilege of sharing a singular space of political discussion, with a reasonable margin for the convergence of ideas, proposals, and actions.

> Given the constituting difference within the RBJA, it's convenient to recognize the diverse way in which their members represent a) the collective they participate in, b) the role of their contribution and c) the nature of the political space in which they expect to act in and its challenges. The understanding of this articulation requires constant problematization and evaluation.[12]

Among the actions unfolded by the RBJA, it is worth mentioning the creation of working groups on different issues that reflect the different interests within the network: environmental racism, chemical industry, mining, and agrobusiness, among others. As it was said before, given its multiple composition, it is difficult for the network as such to have a collective position. Therefore, the formation of working groups means the possibility of working with autonomy and coordination on different issues. Working groups do not only function as a self-education tool, but also as a self-building strategy, as they attempt to create political actors in the public scene.

There are also many solidarity campaigns for emblematic struggles in Brazil. People affected by the use of asbestos, workers exposed to toxic substances, victims of agrobusiness, affected by the big dams, among others, get organized through campaigns. They build strategies of integrated articulation to opposite the strategy of dislocation of big capital in times of economic globalization.

Another important action within the RBJA has been the construction of maps of environmental injustices as collective counter-cartographies. These maps are constructed by the cooperation of affected people, activists and academics, from their experiences and shared analysis, a reappropriation of the knowledge about environmental problems, challenging the state's official discourse.

It is also worth mentioning the project of the Socio-environmental Equity Evaluations (AEA, in its initials in Portuguese) instead of the traditional environmental impact studies in order to incorporate the dimension of the social, labor, and cultural reproduction of the groups affected by the polluting activities. The AEA also questions the government's and corporations' definition of "affected," the obstacles in access to public information, the impossibility of rejecting the polluting projects, the flexibility of the environmental regulations and environment agencies, among others. The project seeks to make the environmental conflicts evident in order to empower democracy.[13]

Finally, it is worth stressing the networking capacity of the RBJA. The participation in world social forums and national encounters is another form in which the RBJA has learned networking networks. One of the present challenges that the RBJA is facing is what they call Dialogues and Convergences towards a Network of Networks, because of the need to construct convergences joining the forces of organized civil society in the fight for social and environmental justice. One main objective is to dispute the political interpretation of the economic, social, environmental, energy, and food crisis, which means to make explicit the real causes of the current crisis and not just a mere explanation of a financial crisis. Other objectives are to communicate the experiences being built by the civil society organizations, to exchange their diverse points of view, and to give public visibility to the achievements and proposals of organized civil society toward another development model in Brazil. Also, to strengthen the capacity of dialogue between the Brazilian networks and society in seeking to overcome the social and political isolation and to make a contribution to revert a political situation described in terms of the fragmentation of the popular struggles in Brazil, the new sociopolitical context, and the criminalization of social movements.

Mexico: The Experience of the National Assembly of Environmentally Affected (Asamblea Nacional de Afectados Ambientales—ANAA)[14]

In Mexico, the signature of the North America Free Trade Area (NAFTA) is the frame that networks fall within. According to different environmental organizations in Mexico, the signature of the NAFTA resulted in the outbreak of dispossession and environmental devastation. This free trade agreement has a competitive advantage for the United States and Canada, such as environmental deregulation and the reduction of the value of the labor force. There are environmental regulations such as those that establish a limit to the pollution of water or the law of ecological balance, but they are not upheld. Environmental devastation has no limits and the Mexican state is considered a failed state, corporative and capitalist, at the same time that there are multiple parallel states—the organized drug cartels, the paramilitary forces.

> At this moment the Mexican state is a state captured by the interests of transnational corporations and organized crime. We have an absolutely failed state under the rule of organized crime. This makes the environmental struggles harder because they are silenced and repressed, the killings and persecutions of the environmental defenders do not stop. Our priority is to have an impact on public opinion, in people's consciousness, in the urban neighborhoods and rural communities.[15]

The ANAA arose in 2008 from this context of multiple community and popular struggles in defense of the natural resources, territory, and people—collective rights. Its original objective was to combat the dispersion and invisibility of environmental injustices. The notion of *affected peoples* in the identity of this network recognizes the consequences of the polluting activities on their bodies, health, and forms of life.

The ANAA started its public activities then with a march to the National Commission of Water and the Environmental Agency (Semarnat) in the context of the World Water Forum (WWF) held in Mexico, where organizations from the entire world gathered to support the Mexican struggles in the first popular discussion on water privatization. They were protesting and demanding from the government a fair solution to the serious problems of pollution of water, land, and air, dispossession of forests and minerals, the permanent deterioration of health, the loss of flora and fauna, and the repression and harassment of communities.

The ANAA's first encounter gathered almost 300 people at the Universidad Autónoma de México. They came from different environmental struggles in Mexico: indigenous communities, peasants' organizations, urban movements, professional associations, academics, civil society organizations, NGOs, just to mention the diversity. Then the assembly started to gain a more organic body, with the formation of working groups and committees. In this context, a discussion about the organizational form took place:

> In Mexico there are organizations for which the structure must be defined by a central committee, that gives directions, that tells the others where they must go, what steps to take. We have been discussing that one big mistake would be to constitute the traditional figure of the political organization back again when problems are so big, when diversity of actors is big and their interests are not the same (. . .) What we wanted was to keep every struggle in its own autonomy, making their own decisions. This is the basic principle of the assembly.[16]

The debate about the forms of political organization recognizes the experience of Zapatismo and the Congreso Nacional Indígena (CNI)—the National Indigenous Congress—in creating open structures for coordination.[17] The ANAA then has implemented a representative council to solve some problems of everyday organization (for instance, the lack of resources for everyone to travel to the meetings and attend all the discussions and agreements). The representative council functions as the assembly political organ in charge of the execution and monitoring of the collective agreements. It also organizes the national encounters and assumes different organizational tasks.

The plenary of the assembly is where deliberation and decision making takes place. The objective is that a general assembly functions with regional assemblies to spread the organization in the country, with tasks of spreading the struggle in the region, self-diagnosis of the environmental, political, and health situation, organizing workshops and other forms of self-education, promoting regional actions, and networking. As Octavio Rosas affirms:

> In our second assembly we had a discussion and we made a decision, that ANAA assemblies would take place in those places where there is a conflict, so people there can receive the solidarity from other struggles, at the center of discussion the local problems, the other *assemblimen and assembliewomen*

> may hear live from the directly affected what that community is doing in self defense and to resist. This is useful to focus on the discussion of an specific problem in which many would find themselves reflected, the assembly has that perspective, that everyone can meet everyone and get to know each other.[18]

Regarding the main actions of the ANAA, it was not until the second encounter that the assemblers started to elaborate a general diagnosis called Environmental Collapse in Mexico. This document denounced the Mexican state because of its responsibility in the destruction of both urban and peasant lives of the Mexican people, through megaprojects such as dams, mines highways, landfills, burial of industrial and toxic waste, privatization of natural resources and public services, and so on.

In this sense, ANAA has developed several actions of self-organization/self-defense/self-education. To start with, we will mention the interdisciplinary clinics where the members of the ANAA deal with the environmental and health problems from different points of view and knowledge. Also, what they call popular and critical epidemiology is pointed out to be an activity that allows local indigenous and peasant struggles to emerge from invisibility and seed public doubt on the official discourse. The practice of a communitarian epidemiology thus is not only reduced to the experts' wisdom. The clinics were useful to rebuild communitarian ties, and also to unify them in their demand for their right to health. Along with this popular and critical epidemiology, it is worth mentioning the making of collective counter-cartographies to map the environmental problems and conflicts.

Also, some initiatives of autonomous environmental impact studies were implemented. While the usual environmental impact studies are designed and applied by experts from government and corporations, the ANAA proved here its capacity to make studies on its own, with the assessment of experts, which is not the same as studies made by experts by themselves and without consulting the population. These so-called "people's (or popular) environmental impact assessments" came mainly from people's own narratives and self-valorization to describe and analyze their situation.

In the same sense, community surveillance committees, with self-proposed participation, are another action in the different nodes of the ANAA network. Given the lack of state controls and the inefficiency or just the complicity of environmental agencies with polluting corporations, affected people make the controls by themselves, documenting the damages and risks, and denouncing them through media, alternative media, and the public space.

Activities of self-education took an important place, such as workshops on issues like the different environmental problems in Mexico, so the affected people could link their own problems to that of others and have a global idea of the causes of these problems. Also important is the training in the use of legal tools, and the capacity of networking.

In this respect, travelling information campaigns about struggles and networks are another important event unfolded by the ANAA, like the process of national and international mobilization to participate in the counter-world summits on climate change, such as the Cancún experience in 2010, along with the Via Campesina. A project of the networks expansion and creation of an international link commission is taking place at the same time as there are other initiatives to spread the network in different scales. We can mention the use of Transnational Ethical Courts that are mechanisms of the global civil society to investigate and denounce human rights violations in Mexico. They also work as an agglutinating frame for organizations, since different struggles find here a space for gathering and put in common their cases to an international audience, including government officials and NGOs. The examples of the Latin American Water Tribunal and the Mexican Chapter of the People's Permanent Court can be mentioned as such spaces that are invoked because citizens can no longer rely on their own national governments, said to be one of the main instigators of the rights violations.[19]

Finally to be highlighted, the connections with academic associations such as the Unión de Científicos Comprometidos con la Sociedad (Union of Scientists Committed to Society) through its environmental observatory has provided not only scientific information but also legitimacy to the ANAA by academics.

Argentina: The Experience of the National Network of Fumigated Peoples (Red Nacional de Pueblos Fumigados)

In Argentina, the use of agrotoxics for the agroindustry—the main economic activity in Argentina—has been denounced because of its harmful effects on health and environment by the affected peoples, the self-called Fumigated Peoples (Pueblos Fumigados). The problem is not only the use of chemical substances but also a context of agriculture modernization and the extension of agrobusiness and production of commodities for the global market. Fully promoted by the national government, the agrobusiness model implies a technical and political subsystem that includes financial capital, chemical industry, biotechnologies and nanotechnologies, and agromachines industries, to name just a few. This model not only implies the displacement of other kinds of agroproduction to transgenic soy monoculture but also the clearing of native forests, the forced eviction of indigenous and peasant communities, and the consequences on health, among other things. Therefore, the Pueblos Fumigados denounce these proactive policies toward agrobusiness and their transfer of costs to the environment, the different forms of life, human and nonhuman, and on biodiversity.

After years of neoliberal government, welfare-state institutions have been dismantled by privatization, decentralization, and budget cuts. In this context, the agencies in charge of regulations have been transformed into mere executors of administrative tasks. The nonexistence or ineffectiveness of neoliberal regulation agencies, which call for individual responsibility instead of protecting rights, has left the population abandoned. In this context, as in 2001 when the citizens threw out the national government, citizens are once again rejecting this transfer of state responsibilities and questioning the state agencies' intervention and legitimacy.

Pueblos Fumigados and many organizations and social movements of civil society started to demand the state structures for these dramatic consequences, particularly the damage to people's health because of the toxins, and also the pollution of soil, water, and even air. The demand on public officials to fulfill their public duties is at the same time the revocation of the legitimacy of their term of office. In this context, the appeal to the rights of citizenship and at the same time the condemnation of new forms of genocide is not a contradiction but a paradox of the fight for rights in Argentina.

> We keep saying that the situation we live in in our neighbourhood is an undisclosed genocide, because we are ending yet another year after more than 10 years complaining, and the economic interests are still more important than people's health. We have laboratory analyses that prove that our children have agrochemicals in their blood, which show that our accusations are not the product of our imagination. The public officials still deny this reality, and therefore, a possible solution
>
> As society we can't celebrate a form of production based on killing, with practices that destroy our neighbourhood relations, community relations, family relations, and our relationship with nature. We have started to build real institutional tools that defend us from the ambition of the agrobusiness producers, and we have the example of other societies that are serious about their health and environment, and we have above all the collective will to make it.[20]

Regarding the organizational form, Pueblos Fumigados have still not constituted a national network or assembly or a national movement. The provinces where the most important soy production is located is where the organized community resists, through local and regional assemblies and regional campaigns that have started at the national level since 2006. In the last two years (2010–2012) there have been increased efforts to reach a national coordination, but there are some obstacles to achieve this goal, such as the capacity to overcome traditional political practices of leadership and neocorporativisms within the social movements in order to create an open and decentralized structure.

Beyond this problem, it is worth highlighting that the assemblies function as spaces of discussion and exchange, and that is where problems are brought up, thought through collectively, and agreements and disagreements about actions, strategies, and plans are formulated. The rejection of traditional political representation (public officials, parties, even bureaucratic unions) is one characteristic, and is also one point discussed on certain occasions ("Should we work with them or not? Should we always keep ourselves separated or not").

The collective value given to the assemblies since the political events of 2001 in Argentina, as spaces for encounter, have achieved the break of the isolation of struggles, but still there are difficulties in network formation. Even though the network is not formalized as such and there are still some obstacles to build a national network, there is low intensity networking that connects experiences through the form of national and regional campaigns, such as the campaign "Stop Spraying Us" (*Paren de Fumigarnos*). Usually the first scope of campaigns is to collect testimonies of the affected people, data about diseases related to the activity of agrochemical spraying, water and soil analysis, among other things. The information goes into a report for public access. It is used for protesting to the national authorities and also for spreading the information to society and communities working on the problem.

One of the processes taking place within the networks is the people's (popular) law-making. The national constitutional principle that people do not govern nor deliberate except through their representatives is subverted by this action. The affected people and professionals committed to them have developed skills, knowledge, and practice in the elaboration of protective regulations against agrotoxics while the official representatives do their legislative work for corporations' interests. Reforms to the existing regulations or new regulations are proposed at a municipal, provincial, or even federal level, with or without success sometimes, but always generating a conflict with the instituted political representatives and the corporation lobbies.

Networks have unfolded different ideas to increase the protection of communities through legislation: prohibit the use of certain agrotoxics, extend the area of exclusion, and so on. They use the instituted People's Initiative (iniciativa popular)[21] and communication campaigns, among other things. This practice faces some obstacles, such as the refusal of the representatives to recognize these constitutional proceedings. Testimonies of the network members show the inequalities in political representation, between the lobby of corporate interests and the guarantee of citizenship rights to life and environment.

In those cases in which networks achieve law reform, there comes another problem: the lack of policy power to ensure its upholding. In this situation, it is the affected people that exercise control through community organization. Alert networks between neighbors have turned into an everyday practice. Community monitoring has permitted the reporting of illegal spraying activities, and also demonstrated the state monitor agencies' inefficacy.

The same methodology of self-organization is applied to community epidemiological surveys and counter-cartographies of the diseases and the polluting problems, given the lack or deficit of the public agencies' interventions.

Finally, the Pueblos Fumigados have also developed their capacity for participating and creating networks geared toward their need to gather and generate genuine spaces of participation and socialization of knowledge and everyone's contributions. Unity is a key question for the spread of transformative political experiences, for the struggle for social change. Not only for the creation of new networks, such as the Red de Médicos de Pueblos Fumigados (a network of medical doctors committed to environmental problems) but also for the learning and continuity of pre-existing networks, a space of convergence of different struggles against extractivism in the region.

Part 2: EJN Common Aspects and Political Power in Latin America

In part two there follows a comparative analysis to highlight the common aspects of these experiences. These aspects will show the ways in which EJN political power against environmental injustices can be thought of as an institutional creativity *from below.*[22]

The notion of institutions is crucial to think about EJN experiences as they can be understood as a set of practices, rules, and habits that come from a constituent power and at the same time, these experimental forms confront and unmake that capital and state sovereignty. Then, as a product of a comparative analysis, three common aspects of this institutional creativity will be presented: first, the forms in which EJN power comes from the exercise of autonomy against the rule of capital and state power; second, the principle of multiplicity and the interaction between local and transnational scales as a key matter for network expansion; and finally, the network's collective experimentation of a diversity of tactics/strategies.

As EJN were described in part one, they arose in a context where state institutions are subsumed under the capitalist command. The defense of life, environment, and health as rights is a task that networks are facing in the crisis of those state institutions/government structures with the public responsibility of rights protection and guarantees. Networks denounce the governments for the exclusion and lack of representation of the citizens' rights, and in doing so, they generate political institutions of another sort.

The critique of political representation is not only the questioning of its legitimacy; it is the building of a form of organization that unfolds the functions that correspond to the state but *without the state.* This means that environmental policies, environmental law and regulation are promoted by the force of networking from below. Also to be highlighted, the practices in the EJN of community epidemiology, popular environmental impact studies, collective cartographies, and self-education workshops, among other things, are common aspects in the Brazilian, Mexican, and Argentinean experiences.

The critique of traditional representation also extends to traditional forms of organization because of its centralized process of decision making, its hierarchical and vertical structures. Instead, the EJN starts from the practice of assembly as a horizontal space and time for encounters and for everyone to speak on behalf of himself or herself without any mediations. However, the exercise of autonomy is about learning from the practical problems of horizontality. How to combine an open structure with decision making, the threat of network collapse, the overdemand of work (too many tasks and too few people) are issues that led to the formation of particular forms of representation such as the councils (Colegiado in Brazil or the Representative Council in Mexico). This form of formal representation in the nonvertical and open structures of networks should not be understood as contradiction but as the challenge of a practical horizontality. Since there is a multiplicity of forms of life and political practices within the networks, the principle of autonomy has also led to the creation of autonomous working groups, such as those in the Brazilian experience, or regional and local assemblies in the Mexican and Argentinean experiences.

Decisions are mostly made in the general assembly, and even when certain positions would not be entirely made in a democratic form, the assembly has always had the power of decision making. The principle of autonomy means the capacity and possibility of every individual and collective to make their own decisions within the organization and its basic agreements. The assemblies will approve and reject the work of their councils and working groups, and they are privileged spaces to define and agree with the rules and proceedings of the network.

Following the comparative analysis, there are other aspects of an institutional creativity from below, beyond horizontality and a self-normative capacity, to be taken into account: the interaction of local and transnational scales as a key matter for network expansion, in a single word—multiplicity. Since this term refers to the power of connections between the singularities (collective, individuals) that come together, it is perhaps the main concern about networking. The EJN have learned about the limits and possibilities of a multiple and open composition—structure based in the mutual understanding of the principles that make multiplicity possible, on the one hand, and the connections of time and space scales, on the other.

The political practice of networks has to do with sharing principles such as diversity, solidarity, trust, and reciprocity, as we have described in the origins of the EJN. The practice of these shared principles allows multiplicity to exist and to expand its power, as much as the lack of it is a loss of power when the principles are not shared. Understanding of diversity as a practice implies also the recognition and speaking of different languages, the dialogue of different knowledges, the encounter of diverse positions and forms of life. The importance of the recognition of differences implies that in cases of conflict there can be place for different positions that may not reach a final agreement but also will not get to the exclusion of some of its members based on the disagreements.

Another lesson from the EJN's practices is the coordination of this multiplicity with the autonomy of each node of the network, as the examples of the working groups and different activities that can be performed without a central command in the EJN proves. Here there is also the valuing of trust and freedom to let everyone work from a basis of common agreements, instead of hierarchical and vertical structures. Acting without constraints, with mutual responsibilities instead of manipulations seems to be one challenge for a practical understanding of multiplicity.

The second issue about multiplicity is what the EJN have learned about scales of space and time, considered as the different levels or dimensions of networking in a context of globalization. The Brazilian experience of RBJA is particularly rich in these matters, since they have learned how space scales in the network's experiences show the limits and possibilities of the interactions between local, national, regional, and global scales. While capital decisions are made at a global scale, its application and effectiveness happens at a local scale. The conflicts are at a local scale but it seems they cannot jump onto the other scale where global decisions are made, where corporations and organisms gather and decide over the life of people and the common goods.

As presented in part one, EJN have different networking capacities. Some of them function just on a global level and miss the connection with the grassroots, where the effectiveness of application happens. A problem pointed out here is the risk of jumping from the local scale to a global one without working out transitions and backups/resonance at a local level, a rearguard for the global. So the network is stronger in its connection and solidarities. The choice for regional assemblies is an attempt to redistribute and guarantee the access to participation and decision making of all the members and not just a few that can always move from one place to another.

Regarding the scales of time, the question about temporality is meaningful when it comes to measure the network power and also its capacity to build agendas and political platforms. The question about whether networks have to last forever is an invitation to think about the power of connections from short term to long term. A network can evaluate its capacity on a short-term scale whether or not it can fulfill its immediate objectives. On a long-term scale, there are other insights such as the network's capacity to build its own frameworks and organizational forms as we have presented in this chapter.

As was said before, the practical understanding of multiplicity and scales implies the possibility to create and expand EJN capacities. The capacity to unfold multiplicity is also linked with the liberation of the networks' creative capacities. Given their multiple composition, another common aspect in the Latin American experience is the diversity of actions that are required by the complexity of the situations that environmental struggles face in the context of globalization.

This diversity of tactics and strategies starts from conceptual creativity. In this regard it is worth mentioning that the notion of environmental justice is more widespread in Brazil rather than in Argentina and Mexico, but in spite of this fact, we believe that the experiences of the ANAA and the Pueblos Fumigados are not far away from the meanings and context in which this notion is used both in the critique of capital and for thinking about another sense of justice that is not reduced to the state's administration of justice. The identity of the affected peoples in Mexico and the references to an environmental genocide in Argentina are concepts that emerge in the context of environmental injustices.

Environmental justice refers to a wide convergence and confluence of movements, collectives, actions, perspectives, and trajectories in the fight for rights. With initiatives based in communities' self-organization and reforms to the state institutions, EJ contributes to the generation of public and common spheres with different tactics and strategies to face the environmental injustices and the hostility and violence of governments and corporations.

In this sense there is no place for a determination or a clear direction on what to do or where to go in the complexity of environmental conflicts. There is an oscillation between innovation and repetition, between creativity and bureaucratization and routines. EJN's actions move away or reinvent traditional forms of protests to the creativity of forms of struggle/forms of life, or reinvent them, give them new life; in other words, the combination of a moment of rupture with a moment of new uses of what already exists. Networking is generated by, and generates, creativity as a collective experimentation. There is a wide arch of tactics and strategies that could be thought of from the regeneration (rethinking, reinventing) of public space institutions, what could be pointed to as reformist positions to more radical positions of exit and creation of new institutions beyond the state.

The first positions could be defined as the regeneration of the public space and refer to those struggles that recognize the state institutions' legitimacy. Nevertheless, this recognition does not imply subjection but the exercise of active demands and an invocation of the democratic state promises. The question to the state institutions does not stop even when it turns out to be a strategic use of the state institutionality, as a mechanism that reveals the state's incapacity to build the public and ensure that the constitutional principles are upheld. The political practices regenerate the public by promoting institutional innovation in the state structure (environmental law-making and democratic procedures to deal with the complexity of environmental problems, participative environmental impact studies).

The experiences that go beyond the public space in the tendency of radical democratic practices and criticisms of state intervention can be described as political practices for which the appeal to the rights of citizenship does not mean subordination but a strengthening of self-organization. The composition of diverse individuals and collectives in the common practice of self-building stems from their own capacity to define ends and means, self-education and reappropriation of knowledge (i.e., community epidemiology), in the complexity of environmental issues. Finally, the capacity to generate a network of networks aims to enforce EJN political power through dialogues and convergences without the intervention of state institutions and agencies.

In summary, the Brazilian, Mexican and Argentinean experiences have common aspects in which the network form of organization means institutional creativity. The practices of autonomy, multiplicity, and diversity shape the possibilities—and limits—of EJN political power to question and restrain the impunity of capital sovereignty and define in the same act experimental ways for radical democratic arrangements to move forward the environmental justice principles.

Conclusion

The presentation situation in Latin America is without doubt marked by several cases of environmental injustices generated by neo-developmentalism and neo-extractivism in the region. Governments have failed to provide recognition and protection to the rights of the affected peoples as they rule under the command of corporations' interests over peoples' rights. This relationship between neodevelopmentalism/neo-extractivism, neoliberalism and environmental injustices in the region is being denounced by EJNs not only from a critical standpoint but from their frameworks and actions. They are creating other forms of political institutions in their claims for rights that put into practice the environmental justice principles here and now.

Institutional creativity is coming from self-organization and the practical understanding of horizontality, multiplicity, and collective experimentation. It could be pointed out that this is not enough evidence to describe the political power of EJN. Still, decision making is a monopoly of governments and corporations, and there is a wide gap between EJN actions and the responses of governments.

The attempts of a global environmental governance as the answer of state and corporation powers to the global demands of climate and environmental justice have not yet reached the generation of real democratic structures to negotiate and reach solutions to the environmental conflicts because they reveal the lack of intention to abandon or even limit the destruction and dispossession of capital. In this context, EJN in Latin America everyday keeps on building these new institutional forms to make lasting and stable a constituent power that calls for the affirmation of the value of life over profit.

Notes

1. I am grateful to the National Council of Scientific and Technological Research of Argentina (CONICET) for providing the financial support to conduct this research. The reflections in this chapter were possible thanks to the cognitive and affective cooperation of the members of: Rede Brasiliera de Justiça Ambiental (Brazil), Asamblea Nacional de Afectados Ambientales and Red en Defensa del Maíz Nativo (Mexico), and the Pueblos Fumigados (Argentina). I would like to mention the permanent advice of Prof. Cecilia Carrizo from my home university, Universidad Nacional de Córdoba, Argentina. Also to be mentioned—the careful and patient edition of this chapter by Adela Alfonsi. Some fragments of a previous version of this work were published in Spanish in my article "Redes en Justicia Ambiental en América Latina. Entre la creatividad democrática y las nuevas instituciones de lo común," in *Justicia Ambiental y Creatividad Democrática,* eds. Carrizo and Berger.

2. Harvey, *The New Imperialism,* 64.

3. Reyes and Kaufman. "Sovereignty, Indigeneity, Territory: Zapatista Autonomy and the New Practices of Decolonization."

4. In terms of my methodological proceedings, the analysis of EJN does not come from the traditional networks analysis as the mapping and measuring of relationships and flows between people, groups, organizations, the nodes and the flows between them. Instead, the research was based in frame analysis and a reconstruction of EJN political practices in their political, institutional, and intersubjective contexts. Interviews, conversations, documents analysis and my personal participation in EJN activities have been the research tools.

5. The frame of political and academic debates about justice, and about environmental justice in particular, give an important place to the organizational and institutional innovation generated by the global movements and citizenship in environmental issues. Some references for these debates that have been a contribution to this chapter can be found in the works of Nancy Fraser, Axel Honneth, and Klaus Eder, among others. Within the environmental justice framework, I would like to thank the academic contributions from David Schlosberg, Kenneth Gould, and Daniel Faber.

6. See RBJA Principles Declaration at http://www.justicaambiental.org.br/_justicaambiental/pagina.php?id=229.

7. Notes from Prof. Henri Acselrad at RBJA web forums.

8. Firpo, Movements and the Network of Environmental Justice in Brazil, 101. See also the introduction of Acselrad et al., *Justica Ambiental e Cidadania.*

9. Ibid.

10. http://www.justicaambiental.org.br/_justicaambiental/pagina.php?id=912

11. Words of Marcelo Calasanz in the IV National Encounter of the RBJA. Records by author.

12. Ibid.

13. http://www.justicaambiental.org.br/projetos/clientes/noar/noar/UserFiles/17/File/Encarte_AEA_2ed.pdf

14. http://www.afectadosambientales.org/

15. Interview by author with Octavio Rosas, Asamblea Nacional de Afectados Ambientales (October 2011).

16. Ibid.

17. As Ana de Ita from the Network in Defense of the Native Maize (Red en Defensa del Maíz Nativo) says: "Many of us who were participating in the Congress were also participating in the Zapatista Movement and the San Andres Dialogues on Indigenous Rights in Mexico. One of the main debates was to build an organization or not, the Zapatistas and the Indigenous said they would just form a network, because nobody could speak on behalf of the National Indigenous Congress, because this congress exists only when everybody would gather, and that they would not delegate any responsibility in a fixed structure. If they were not together they would act as independent organizations, and when they gather they would function as a network, and this idea is still present in many peoples." Interview by author (October 2011).

18. Interview by author with Octavio Rosas, Asamblea Nacional de Afectados Ambientales (October 2011).

19. Check the Latin American Water Tribunal website: http://www.tragua.com/index_english.html and The People's Permanent Court—Mexican Chapter (Tribunal Permanente de los Pueblos): http://www.tppmexico.org/

20. First quote is from the Group of Mothers of Ituzaingo Neighbourhood, the second quote is from the Collective Paren de Fumigar (Stop the Sprayings) from the province of Córdoba, Argentina. To read about the use of the notion of Environmental Genocide see C. Carrizo and M. Berger, *Citizens' Rights and Environmental Genocide.*

21. Legal instrument instituted in the Constitution consisting in the faculty of the citizens to propose law reforms or initiatives. This instrument is currently not used due to the lack of legitimacy of the state institutions, but people have learned to politicize these tools and make use of them not only as a legal proceeding but also as a way to enforce the public debates.

22. The idea of institutional creativity from below is under construction with different but convergent contributions. From John Dewey and Hans Joas's ideas of creative democracy to Hardt and Negri's ideas of institutions of the commons, Paolo Virno's nonstate public spheres, and Maurizzio Lazzarato's nomad coordinations, among others.

Bibliography

Acselrad, H., J. Padua, and S. Herculano, eds. *Justiça Ambiental e Cidadania.* Rio de Janeiro: Relume Dumará, 2004.

Berger, M. "Redes en Justicia Ambiental en América Latina. Entre la creatividad democrática y las nuevas instituciones de lo común." In *Justicia Ambiental y Creatividad Democrática,* edited by C. Carrizo and M. Berger. Córdoba: Alción Editora, 2012.

Carrizo, C., and M. Berger. "Citizen's Rights and Environmental Genocide." *Environmental Justice Journal* 5, no. 2 (2012): 105–10.

Dewey, J. *The Public and its Problems.* New York: Ohio University Press, 1954 (1927).

Eder, K. *The Social Construction of Nature.* London: Sage Publications, 1996.

Faber, D. "Building a Transnational Environmental Justice Movement: Obstacles and Opportunities in the Age of Globalization." In *Coalitions across Borders: Negotiating Difference and Unity in Transnational Struggles against Neoliberalism,* edited by Joe Bandy and Jackie Smith. New York: Roman & Littlefield, 2005.

Firpo, M. "Movements and the Network of Environmental Justice in Brazil." *Environmental Justice Journal* 5, no. 2 (2012): 100–04.

Fraser, N. *Scales of Justice: Reimagining Political Space in a Globalizing World.* New York: Columbia University Press, 2009.

Gould, K., D. Pellow, and A. Schnaiberg. *The Treadmill of Production: Injustice and Unsustainability in the Global Economy.* Boulder, CO: Paradigm Publishers, 2008.

Hardt, M., and A. Negri. *Commonwealth.* Cambridge: Harvard University Press, 2009.

Hardt, M., and A. Negri. *Multitude. War and Democracy in the Age of Empire.* New York: Penguin, 2004.

Harvey, D. *The New Imperialism.* New York: Oxford University Press, 2003.

Honneth, A. *The Struggle for Recognition: The Moral Grammar of Social Conflicts.* Cambridge, MA: MIT Press, 1995.

Joas, H. *The Creativity of Action.* Chicago: The University of Chicago Press, 1996.

Lazzaratto, M. *Por una política menor: política y acontecimiento en las sociedades de control.* Madrid: Traficantes de Sueños, 2006.

Reyes, A., and M. Kaufman. "Sovereignty, Indigeneity, Territory: Zapatista Autonomy and the New Practices of Decolonization." *The South Atlantic Quarterly* 110, no. 2 (2011): 505–25.

Schlosberg, D. "Networks and Mobile Arrangements: Organisational Innovation in the US Environmental Justice Movement." *Environmental Politics* 8, no. 1 (1995): 122–48.

Schlosberg, D. "Reconceiving Environmental Justice: Global Movements and Political Theories." *Environmental Politics* 13, no. 3 (2004): 517–40.

Virno, P. *A Grammar of the Multitude: For an Analysis of Contemporary Forms of Life.* New York: Semiotext(e), 2004.

PART II

North America

Chapter 6

Environmental Movement in Canada

Laurie Adkin

Canada is a big country and its environmental movement is correspondingly diverse—shaped significantly by the country's geography and federal political system. Growing out of local geographies and economies, the issues that animate environmentalists in Canada vary to some extent from region to region. The organizational structure of the movement and its resources for shaping environmental policies are greatly influenced by the country's federal system as well as by the treaty system. So, to understand the particular morphology of the Canadian environmental movement, we need to start with some of the basic features of Canada's geography and political system.[1]

The Physical Geography of Canadian Environmentalism

Canada covers half a continent (9,984,670 square kilometers), making it the world's second largest country after Russia. Its border with the United States is 8,890 kilometers long, and most of the population lives in the southern part of the country (concentrated in the provinces of Quebec and Ontario and in the largest provincial cities) (see Table 6.1).

However, approximately 40 percent of Canada's landmass and freshwater is located north of the 60 degrees North latitude. With their enormous lakes and rivers, the Northwest Territories and the Territory of Nunavut alone contain 9.2 percent of the world's total freshwater.[2] It is estimated that Canadian rivers discharge close to 9 percent of the world's renewable water supply. Canada has more than 31,750 lakes—so many that the exact number is unknown.[3] The Great Lakes—Superior, Michigan (located entirely in the United States), Huron, Erie, and Ontario—are the largest group of freshwater lakes in the world. In addition to all this fresh water, Canada has the world's longest maritime coastline (202,080 kilometers) when the east, west, and north coasts are added up.

Canada is also heavily forested (approximately 45 percent) compared with many other parts of the world.[4] Canada's recoverable oil reserves—second only to Saudi Arabia's—make this country (and especially the province of Alberta) a major exporter of crude oil to the United States and add to the Canadians' large per capita contributions to global greenhouse gas emissions. In sum, Canada is a land rich in fresh water, maritime access, forests, and mineral resources, while having a very small proportion of the world's human population (less than 1 percent) and a potentially large role to play in creating solutions to the planet's urgent ecological problems.

The Political Geography of Canadian Environmentalism

In their efforts to bring about regulatory or policy changes, environmentalists in Canada have to navigate the division of jurisdictional responsibilities among municipal, provincial, and federal levels of government. Canada has 10 provinces and 3 territories (Yukon, Northwest Territories, Nunavut), and the Constitution Act of 1867 assigned different responsibilities to these levels of government. This means that in some areas (e.g., fisheries, navigable waters, general "peace, order, and good government," criminal law, trade, foreign policy, aboriginal peoples) the federal government exercises legal jurisdiction whereas in other areas (e.g., ownership of public lands and natural resources, property and civil rights, determination of the authorities granted to municipal governments, land use, labor law, education) the provinces have jurisdiction. Responsibility for environmental protection is not explicitly assigned by the Constitution Act; environmental policy—along with water, public health, and agriculture—is an area of shared jurisdiction because it is crisscrossed by responsibilities held by different levels of government. For example, both levels of government have responsibilities for conducting environmental assessments. Under the Canadian Environmental Protection Act (1988), the federal government may classify particular substances as toxic chemicals and regulate their use. However, the subnational governments also have their own environment Acts that regulate various pollutants. The Canadian Water Act requires the federal government to try to make agreements with any provinces concerned before acting on water quality issues. The 2002 federal Species at Risk Act relies for implementation upon a council made up of ministers responsible for wildlife at the federal, provincial, and territorial levels.

Indeed, federal legislation that touches upon provincial jurisdiction over natural resources has, as a rule, been implemented through consultation with provincial governments and has not necessarily created national (uniform) standards. In turn, this means that federal environmental legislation is also negotiated with private sector groups (e.g., the mining industry, or the oil and gas industry) that exercise substantial influence over provincial governments. The absence of a Canadian Environmental Bill of Rights and the reluctance of federal governments to assert (or test) their constitutional powers to legislate in the area of environmental protection have contributed to what is described as a "patchwork" of environmental laws and standards across the country. Overall, the provinces generally dominate in the area of environmental policy-making—a situation that was reinforced in 1999 by negotiation of the Canada-wide Accord on Environmental Harmonization, as a result of which the federal environment minister undertook "not to act" once certain environmental responsibilities have been assigned to the provinces.[5]

Further complicating environmental policy-making processes in Canada are the relations between the federal, provincial, and aboriginal governments. Canadian courts have determined that governments have a constitutional "duty to consult" aboriginal peoples when a proposed "development" project is likely to infringe upon Aboriginal or Treaty rights.[6] Aboriginal rights constitute resources that may be brought to bear in conflicts in which environmentalists and other actors are also engaged. However, rather like the "triggers" imbedded in environmental impact assessment laws, such rights may be ruled inoperative by the courts, and do not constitute a veto power vis-à-vis proposed "development."

Finally, in the context of failures at provincial, national, and international levels to act on such urgent issues as climate change, growing attention is being paid to the roles for municipal governments in implementing ecologically sustainable development. Local governments have direct, shared, or indirect control or influence over land use planning, waste management, water and wastewater supply and treatment, building codes, public transit, and other activities.

These jurisdictional questions matter to environmentalists because they structure the scope and foci of environmental campaigns, and because they determine the juridical resources and policy tools available to environmentalists and other actors in any particular conflict. Can the federal government, for example, be called upon to intervene in the approval process for an industrial project (e.g., a mine, a waste landfill, a hydroelectric dam, or an oil refinery)? This depends upon a number of considerations, such as on the interpretation of the federal Environmental Assessment and Review Process rules (1984), or on whether or not fish habitat will be affected (a Canadian Fisheries and Oceans responsibility). On the other hand, municipal governments are playing an increasingly important role in opposing projects subject to federal approval (e.g., by the National Energy Board), or legislative changes affecting environmental protections (e.g., amendments to environmental legislation such as the Fisheries Act in the federal Bill C-38, introduced in spring 2012).

In addition to the complexities of the Constitution Act, The Canadian Charter of Rights and Freedoms (1982) contains several clauses that may constitute environmental or health rights. For example, some environmental lawyers have argued that section 7 of the Charter, guaranteeing "the right to life, liberty, and security of the person," may be used to make the case that Canadian First Nations have a constitutional right to clean water.[7] Constitutional rights are, in general, an important resource for citizens engaged in environmental conflicts.

Mapping the Canadian Environmental Movement by Region and Issue

Because of Canada's federal political system, the environmental movement organizes itself locally, provincially, and federally, as well as in issue-based, cross-country coalitions. Most provinces (and one territory—Yukon) have environmental networks that try to coordinate their members' activities and sharing of resources. At the same time, all of the provincial networks are linked to a national network: the Canadian Environmental Network (RCEN).[8] In 2011, the RCEN reported that it encompassed more than 620 associations across the country, and a membership of at least 630,000 individuals directly involved through their groups (see Table 6.1).[9]

Language still constitutes a substantial barrier to awareness in one part of the country about what is going on in other parts of the country (whether with regard to politics, culture, or social movement campaigns). However, within the predominantly francophone province of Quebec, there is an environmental network (the SOEQ) that participates in the RCEN and that constitutes the main link between environmental organizations within Quebec and those in predominantly English-speaking Canada. In addition, some of the large ENGOs (like Sierra Club, Greenpeace, and CPAWS) have branches in Quebec, and environmental activists exchange information

Table 6.1. Canadian Environmental Network/ Réseau canadien de l'environnement

Provincial or Territorial Network[a]	Date Established	Number of Affiliated Organizations as of March 31, 2011[b]	Provincial Population as of July 1, 2011 (thousands)	ENGOs per 100,000 Persons
Alberta Environmental Network	1980	77	3,779.4	2.0
British Columbia Environmental Network	1981	62	4,573.3	1.4
Manitoba Eco-Network	1983	56	1,250.6	4.5
New Brunswick Environmental Network	1991	69	755.5	9.1
Newfoundland and Labrador Environmental Network	1990	22	510.6	4.3
Nova Scotia Environmental Network	1991	54	945.4	5.7
Ontario Environmental Network	1981 (1987)[d]	126[c]	13,373.0	0.9
Prince Edward Island Eco-Net	1991	29	145.9	19.9
Le Secrétariat des organismes environnementaux du Québec (SOEQ)[e]	2007	62	7,979.7	0.8
Saskatchewan Eco-Network	1993	56	1,057.9	5.3

(*Continued*)

Table 6.1. (*Continued*)

Provincial or Territorial Network[a]	Date Established	Number of Affiliated Organizations as of March 31, 2011[b]	Provincial Population as of July 1, 2011 (thousands)	ENGOs per 100,000 Persons
Yukon Environmental Network	c. 1986	13	34.7	37.5
CEN/RCEN membership as of March 31, 2011	1977	626	(total) 34,482.8	1.8

[a] The Northwest Territories and Nunavut do not have environmental networks at this time.

[b] The source for this is the most current list of members for each network as of October 2011. Note that these numbers exclude many citizens' groups that are not affiliated with the provincial networks.

[c] As of June 2012, the OEN's website lists 191 "members and subscribers" and states that Ontario has over 500 environmental groups from citizens' groups to branches of national organizations (http://oen.ca/).

[d] In July 1981, the founding meeting of the Ontario ENGO Network was held in New Liskeard. In 1987 it was incorporated as the OEN.

[e] The SOEQ was formerly the Reseau affilié des associations environnementales du Québec (RAAEQ), founded March 10, 2007 (http://credelaval.qc.ca/downloads/vol1no2.pdf).

Sources: RCEN Annual Report, 2010–2011, p. 5, http://rcen.ca/sites/default/files/uploads/2010-2011_rcen_annual_report.pdf; websites of the respective provincial or territorial networks; personal communications with staff from OEN, SOEQ, and YEN, October 30–November 1, 2011; Statistics Canada, CANSIM, table 051-0001, Population by year, by province and territory, http://www40.statcan.ca/l01/cst01/demo02a-eng.htm.

with their counterparts in other provinces where similar campaigns may be taking place. Environmental groups in Quebec are certainly concerned with all of the same issues that animate environmentalists elsewhere in the country. One example is opposition to "fracking"—the hydraulic fracturing of rock or shale formations for the extraction of natural gas. However—as we see in the next section—there are some significant differences as well, stemming from the unique character of civil society in Quebec. Organizationally, there are 15 regional environmental councils in Quebec, as well as numerous ENGOs, citizens' groups, and coalitions. Many of these are listed in Table 6.3 (ENGOs in Quebec).

Provincial and national networks do work together on many issues—particularly those of national and global scale—such as toxic chemicals regulation (a federal responsibility under the Canadian Environmental Protection Act), or climate change. (Greenhouse gases have been declared substances subject to regulation under the Canadian Environmental Protection Act (CEPA), and the federal government ratified the Kyoto Protocol in 2002.) Well-known, nongovernmental, environmental organizations (ENGOs) with national or regional offices and staffs belong to such networks, along with smaller, more local organizations.[10] In addition, labor organizations (trade union local committees, regional labor councils, or provincial labor federations) participate in some of the provincial environmental networks, as do a number of health associations (like the Canadian Lung Association).

Local environmental associations are typically run by volunteers, while the ENGOs generally have paid staff and are funded by member donations or fees.[11] Since 1977, the Canadian Environmental Network has received operating grants from Environment Canada (EC), which have allowed the RCEN to maintain an office and to support the provincial networks to do the same. This 34-year relationship with EC was abruptly ended, however, in October 2011, when the recently elected Conservative government eliminated EC funding for the RCEN. As a result, the RCEN and the 11 provincial and territorial networks dependent upon funding from the RCEN to maintain their offices were forced to lay off staff. Think tanks like the Pembina Institute receive funding from individual donations, foundation grants, and consulting fees for commissioned research. For many years the RCEN had an umbrella funding agreement with Environment Canada for its organization of ENGO stakeholder representation in government consultation processes.[12]

While Table 6.1 provides a picture of the regional distribution of environmental organizations, Tables 6.2 and 6.3 provide a list of the country's major ENGOs. These are typically headquartered in a large urban center (Toronto, Montreal, Vancouver, or Ottawa, the national capital), and have regional offices. While many of these ENGOs engage in a variety of campaigns, some are quite focused on particular issues.

Some ENGOs and think tanks utilize what is called an "insider" strategy for influencing government policy. That is, they provide expert knowledge and participate in consultation processes at various levels with government policy-makers. The mobilization of media campaigns and the organization of collective actions to increase public awareness of issues, or to shape public opinion and to pressure government actors, is work done by a plethora of ENGOs, citizens' groups, and coalitions across the country.

Table 6.2. Major ENGOs in Canada outside of Quebec

ENGO Operating on a National Scale in Canada	Date Established in Canada	Key Functions or Issues
Ducks Unlimited Canada Corporate headquarters in Stonewall, Manitoba	1938	Scientific research and public education aimed at wetland conservation; protection of duck species' habitats for hunters; partnerships with private landowners to protect habitat http://www.ducks.ca/
Bird Studies Canada Headquarters and national research centre in Port Rowan, ON, and seven regional offices Originally founded as the Long Point Bird Observatory in 1960	1960	Dedicated to the study and conservation of wild birds and their habitats http://www.bsc-eoc.org/
Canadian Wildlife Federation (300,000 members)	1962	Protect wild species and habitats http://www.cwf-fcf.org/en/index.html
Nature Conservancy of Canada Headquarters in Toronto	1962	Seeks to protect "ecologically significant lands" through the purchase, donation, or placing of conservation easements. http://www.natureconservancy.ca/site/PageServer?pagename=ncc_main
Canadian Parks and Wilderness Society (13 chapters; 50 staff; 40,000 supporters)	1963	Research, advocacy, public education to protect Canada's wild ecosystems and species in parks, wilderness, and similar natural areas http://www.cpaws.org/
World Wildlife Fund Canada (150,000 supporters) Head office in Toronto; regional offices in Halifax, Montreal, Ottawa, Prince Rupert, St. Albert, St. John's, Vancouver	1967	Wildlife conservation, climate, and energy alternatives http://wwf.ca/

(*Continued*)

Table 6.2. (*Continued*)

ENGO Operating on a National Scale in Canada	Date Established in Canada	Key Functions or Issues
Pollution Probe Office in Toronto	1969	Policy analysis, scientific research, public education, with focus on air and transportation, energy, climate change, and water issues http://www.pollution probe.org/
Canadian Environmental Law Association Office in Toronto	c. 1970	Research, legal interventions http://www.cela.ca/
Greenpeace Canada Main office in Toronto; regional offices in Edmonton, Montreal, Vancouver (opened at different dates) 89,000 supporters in Canada	1971	"Greenpeace is an independent, campaigning organisation which uses non-violent, creative confrontation to expose global environmental problems, and to force the solutions which are essential to a green and peaceful future." http://www.green peace.org/canada/en/splash/
Canadian Environmental Network / Reseau canadien de l'environnement (626 member organizations) Office in Ottawa	1977	Co-ordinating body for national environmental organizations http://rcen.ca/home
Friends of the Earth Canada/ Ami(e)s de la Terre Office in Ottawa	1978	Water, environmental justice, climate change http://www .foecanada.org/
Canadian Coalition for Nuclear Responsibility/ Regroupement pour la surveillance du nucléaire (based in Montreal)	1978	Research, public education, advocacy in opposition to uranium mining, nuclear energy, and nuclear weapons www.ccnr.org
Energy Probe Research Foundation (separated from Pollution Probe in 1980). Its divisions include:	1981	http://epresearchfoundation .wordpress.com/ Energy Probe: Research and advocacy on energy issues, esp. nuclear power http:// ep.probeinternational.org/

(*Continued*)

Table 6.2. (*Continued*)

ENGO Operating on a National Scale in Canada	Date Established in Canada	Key Functions or Issues
Energy Probe, Probe International, Environment Probe, and the Urban Renaissance Institute. Toronto-based.		Environment Probe (1989): policy analysis http://environment.probeinternational.org/ Probe International: http://journal.probeinternational.org/ Urban Renaissance Institute: http://urbanrenaissance.probeinternational.org/
Great Lakes United (binational[a])	1983	Research, advocacy on environmental issues affecting the Great Lakes Basin http://www.glu.org/en/splash
Environmental Defence Offices in Toronto and Ottawa	1984	Campaigning regarding multiple issues http://environmentaldefence.ca/about
Pembina Institute for Alternative Development founded in Drayton Valley, Alberta; offices in Drayton Valley, Calgary, Edmonton, Ottawa, Toronto, Vancouver, Yellowknife	1985	Research, consulting, advocacy, on energy issues and sustainable development www.pembina.org
David Suzuki Foundation Office in Vancouver Staff of about 45; 25,000 donors	1990	Research, public education, advocacy http://www.davidsuzuki.org/
Indigenous Environmental Network (binational) Main office in Bemidji, MN (USA); 1 Canadian campaigner, based in Ottawa	1990	Educate and empower indigenous peoples; opposition to Alberta tar sands and oil pipelines http://www.ienearth.org/
Ecojustice (founded in 1990 as the Sierra Legal Defence Fund; renamed EcoJustice in September 2007) (30,000 supporters)	1990	Legal interventions, science research http://www.ecojustice.ca/

(*Continued*)

Table 6.2. (*Continued*)

ENGO Operating on a National Scale in Canada	Date Established in Canada	Key Functions or Issues
International Institute for Sustainable Development (IISS) head office in Winnipeg, Manitoba and branches in Ottawa, Ontario; New York, NY; and Geneva, Switzerland over 100 staff and associates located in over 30 countries.	1990	IISD receives core operating support from the Government of Canada, as provided through the Canadian International Development Agency, the International Development Research Centre, and the Province of Manitoba. The institute also receives project funding from numerous governments inside and outside Canada, United Nations agencies, foundations and the private sector. http://www.iisd.org/
Sierra Club Canada National office in Ottawa; Regional offices in Halifax, Victoria, Montreal, Toronto, Edmonton, and 18 local groups.	1992[b]	Health and environment, protecting biodiversity, atmosphere and energy, transition to a sustainable economy http://www.sierraclub.ca/
Green Communities Canada (40 member organizations) Head office in Peterborough, ON	1996	Provides information and services to individuals and communities to improve energy efficiency and support sustainable lifestyles http://greencommunitiescanada.org/
MiningWatch Canada/ Mines Alerte Office in Ottawa	1999	Monitors the activities of mining companies operating in Canada, Canadian mining companies operating in other countries, and corporate accountability http://www.miningwatch.ca/
Green Budget Coalition (20 ENGOs)	1999	Prepares an annual "green" budget for Canada and evaluates federal budgets http://www.greenbudget.ca/main_e.html

(*Continued*)

Table 6.2. (*Continued*)

ENGO Operating on a National Scale in Canada	Date Established in Canada	Key Functions or Issues
Forest Ethics (binational) Canadian office in Vancouver	2000	Protection of forests through market solutions http://forestethics.org/
Climate Action Network/Réseau action climat Canada (75 organizations) Office in Montreal	c. 2003	http://www.climateactionnetwork.ca/
Canadian Renewable Energy Alliance (alliance of 16 ENGOs) (executive officers based in different cities)	c. 2004	Transition to renewable energies http://www.canrea.ca/site/
Canadian Youth Climate Coalition (48 youth organizations)	2006	Education work in schools, media; Climate Justice Tribunal, Nov. 2011, Ottawa http://www.ourclimate.ca/wordpress/

[a] GLW has ceased operation as of July 1, 2013.

[b] Regional branches affiliated with Sierra Club International have been operating in Canada since 1963; the national office opened in 1989, and Sierra Club Canada incorporated in 1992.

Sources: Websites for the respective organizations and e-mail correspondence with staff.

Table 6.3. Major ENGOs in Quebec

STOP	1970	Ecological Citizens' Group Incorporated in 1970
ENvironnement JEUnesse	1979	"ENvironnement JEUnesse stimule le développement d'une conscience environnementale et d'une pensée critique auprès des jeunes afin qu'ils exercent des actions citoyennes pour un avenir viable." http://enjeu.qc.ca/
Nature-Québec	1981	Travaillant au maintien de la diversité des espèces et des écosystèmes, Nature Québec souscrit aux objectifs de la Stratégie mondiale de conservation de l'Union mondiale pour la nature (UICN). www.naturequebec.org
Association québécoise de lutte contre la pollution atmospherique	1982	"L'Association québécoise de lutte contre la pollution atmosphérique (AQLPA) s'est donnée pour mission de contribuer à la protection de l'air et de l'atmosphère, à la fois pour

(*Continued*)

Table 6.3. (*Continued*)

		la santé des humains et des écosystèmes." http://www.aqlpa.com/
Réseau Québécois des groups écologists (RQGE)	c. 1983	"Le RQGE favorise la mise en commun des connaissances et des actions des membres. Le RQGE représente les groupes dans leurs revendications communes auprès du gouvernement, notamment concernant une reconnaissance concrète et un financement adéquat du mouvement environnemental pour ses apports à la protection du patrimoine naturel, à la santé publique et au mieux-être de la société québécoise." http://www.rqge.qc.ca/content/historique
Nature-Action Québec	1986	Influencer la société en faisant la promotion de nouvelles façons de faire plus harmonieuses. Celles-ci garantissent à l'homme d'aujourd'hui et à celui de demain un environnement sain et durable où nature et humanité ne sont plus séparées. www.nature-action.qc.ca
Fondation québécoise en environnement	1987	Provoquer et accélérer le changement des attitudes et des habitudes à l'égard de la mise en valeur et de la protection de l'environnement, informer et sensibiliser la population sur les enjeux environnementaux, éduquer la population aux moyens et actions à entreprendre, favoriser la recherche et la mise en place de solutions concrètes et efficaces. www.fqe.qc.ca
Attention Fragile	1988	Protection des Iles Magdalen http://www.attentionfragiles.org/
Centre Québécois du droit de l'environnement	1989	"Depuis sa fondation, le CQDE dispense de l'information juridique à des citoyens et des groupes de protection de l'environnement, leur permettant de faire la lumière sur les dimensions juridiques des problèmes environnementaux auxquels ils font face." http://www.cqde.org/
Regroupement national des conseils régionaux de l'environnement du Québec	1991	Être le porte-parole des orientations communes des régions, assumer un rôle de concertation, d'animation et d'information et offrir des ressources et un soutien aux seize conseils régionaux de l'environnement. www.rncreq.org

(*Continued*)

Table 6.3. (*Continued*)

Front commun québécois pour une gestion écologique des déchets	1992	Mettre sur pied d'autres choix aux méthodes traditionnelles de traitement des déchets. Nouvelles méthodes fondées sur une approche écologique et démocratique: réduction à la source, réutilisation, recyclage / compostage. Éducation, sensibilisation, mobilisation dans le domaine de la gestion écologique des déchets. www.fcqged.org
Coalition jeunesse Sierra	1996	Impliquer les Canadiens et Canadiennes de 14 à 26 ans dans la lutte pour développer un Canada écologiquement et socialement durable. Un projet majeur au Québec : Campus durables. Réseau d'étudiants, 15 universités québécoises. www.syc-cjs.org
Eau Secours! (coalition)	1997	"Le but de la coalition est de revendiquer et promouvoir une gestion responsable de l'eau dans une perspective de santé publique, d'équité, d'accessibilité, de défense collective des droits de la population, d'amélioration des compétences citoyennes des citoyens, de développement durable et de souveraineté collective sur cette ressource vitale et stratégique." http://eausecours.org/qui/
Équiterre Offices in Montréal and Québec City	1998[a]	Objectives: to build a social movement to lead individuals, organizations, and governments, to adopt choices that are ecological, egalitarian, and solidaristic. Works in multiple areas, including climate change. http://www.equiterre.org/
Réseau québécois des femmes en environnement	1999/ 2000	Protéger l'environnement et la santé. Favoriser les liens entre les femmes intéressées par l'environnement. Renforcer les capacités d'intervention des femmes en environnement. www.rqfe.org
Regroupement des organismes de bassins versants du Québec	2001	Rassembler les organismes de bassins versants du Québec afin de favoriser la gouvernance de l'eau dans le cadre de la gestion intégrée de l'eau par bassin versant dans une perspective de développement durable. www.robvq.qc

(*Continued*)

Table 6.3. (*Continued*)

Société pour la nature et les parcs du Canada	2001	Organisme travaillant pour la protection de la nature, favorisant la création d'aires protégées ainsi que l'amélioration de la gestion des aires existantes. Le SNAP désire protéger les grandes superficies de territoires publics. www.snapqc.org
Centre de l'environnement—Écobâtiment	2002	Soutenir les groupes environnementaux dans leur lutte pour la défense du droit à un environnement de qualité pour tous; contribuer à l'atteinte du développement durable par la démonstration, l'expérimentation et l'exhibition d'écotechnologies; fournir des services dédiés au développement du bâtiment durable au Québec et ailleurs dans le monde, etc. www.centreenvironnement.org
Fondations Rivières	2002	"La Fondation Rivières est un organisme à but non lucratif dont la mission est d'oeuvrer à la préservation, la restauration et la mise en valeur du caractère naturel des rivières—tout autant que de la qualité de l'eau." http://fondationrivieres.org/?lang=en
Sierra Club Québec	2003	Protéger les écosystèmes de la planète, biodiversité, éco-santé, économie durable, changements climatiques. www.quebec.sierraclub.ca
Réseau québécois pour la simplicité volontaire	2003	Réunir des personnes qui veulent vivre et promouvoir la simplicité volontaire comme moyen d'améliorer leur propre vie et contribuer à édifier une société plus juste et durable. www.simplicitevolontaire.org
Mouvement Québécois pour une Décroissance Conviviale	2007	-a group that promotes social ecology "Deux objectifs principaux: • Provoquer dans la population du Québec la prise de conscience de l'impossibilité de poursuivre la croissance économique. • Travailler à la mise en place d'une société équitable, autonome, solidaire et frugale." http://www.decroissance.qc.ca/
Mouvement Sortons le Québec du Nucléaire	2008	-anti-nuclear movement "Objectifs: -sensibiliser et informer les citoyens et les décideurs politiques des tenants et

(*Continued*)

Table 6.3. (*Continued*)

		aboutissants du projet de réfection de la centrale nucléaire de Gentilly-2; -mobiliser la population québécoise sur cet important enjeu, en développant un débat de société indispensable, solidaire et rassembleur." http://www.sortonsquebecnucleaire.org/index.php
Pour que Québec ait meilleure mine	2008	Group opposed to uranium mining in Quebec "La coalition a pour mission d'engager le Québec dans une réforme en profondeur de l'exploration et de l'exploitation de ses ressources minérales afin de permettre le développement futur du secteur tout en assurant une protection accrue de l'environnement et de la qualité de vie des citoyens." http://www.quebecmeilleuremine.org/
Maîtres Chez Nous, 21ème Siècle (MCN21)	2009	"Le but est . . . de mener à bout le projet d'indépendance énergétique du Québec de toute forme d'énergie polluante, importée ou non." http://mcn21.org/publications/declaration-maitres-chez-nous/
Mobilisation Gaz de schiste de Saint-Marc-sur-Richelieu et de la Vallée du Saint-Laurent	2010	-regroupement citoyen "Nous demandons un moratoire sur l'exploration et l'exploitation des gaz de schiste qui menacent la qualité des milieux de vie et la sécurité des populations. Nous souhaitons que le gouvernement encourage un système de production et de consommation d'énergies socialement et écologiquement responsable." http://mobilisationgazdeschiste.blogspot.com/ -In 2010 it formed the Comité interrégional gaz de schiste de la Vallée du St-Laurent with other citizens' groups
Comité interrégional gaz de schiste de la Vallée du St-Laurent	2010	Some of the objectives are: -"acquérir et diffuser l'information pertinente concernant les gaz de schiste afin d'encourager et de soutenir la mobilisation citoyenne par l'intermédiaire des comités de citoyens régionaux et locaux;" -"favorises l'unification des forces pour combattre le développement risqué et anarchique de l'industrie des gaz de schiste au Québec;"

(*Continued*)

Table 6.3. (*Continued*)

		-"proposer des actions nationales en faveur de: • l'imposition d'un moratoire générique; • la réalisation d'études indépendantes et scientifiques sur les impacts socio-sanitaires; économiques et environnementaux; • la réalisation d'études indépendantes et scientifiques concernant le développement des gaz de schiste en comparaison d'énergies dites alternatives; • l'organisation d'un débat sociétal concernant les futurs choix énergétiques du Québec." http://regroupementgazdeschiste.com/2/index.html
Mouvement Vigilance Energie (coalition)	2010	network of 14 ENGOs and citizens' groups "Mission: Assurer une veille stratégique permanente sur les questions d'énergie au Québec tant au niveau écologique, économique et sociologique en plus de faciliter la mise en commun d'expertises et de ressources sur les questions de l'énergie au Québec." http://www.mcn21.org/communiques/article/lancement-du-mouvement-vigilance

[a] Équiterre has its origins in the Rio Summit of 1992. It was incorporated as ASEED in 1995 and changed its name to Équiterre in 1998.

Sources: Websites of the organizations; e-mail correspondence with ENGO staff; Nature Québec et STOP (2010).

In addition to the local citizens' groups and the more institutionalized environmental organizations that one might see as constituting the Canadian environmental movement, we should acknowledge the important resources provided to citizen activists by researchers based in universities or nonprofit institutes. As problems such as climate change, loss of biodiversity, or environmental toxins grow more pressing, conservation biologists and earth scientists, in particular, are playing increasingly important roles in bringing their research to the public, providing expert testimony in various public consultation processes, and policy advocacy. Social science researchers, too, are providing policy analysis and critiques of existing decision making processes, as well as advocating reforms to democratize institutions and green our political economy. Outside of the universities, research is produced by think tanks like the Pembina Institute, the David Suzuki Foundation, the Canadian Centre for Policy Alternatives, Parkland Institute, Polaris Institute, or the Great Lakes Institute for Environmental Research. Of course, research relevant to ecological transition is emerging from many other locations as well, including unions, consultants hired by communities to investigate their concerns, and ENGOs like Greenpeace, Ecojustice, the Canadian Environmental Law Association, the Association québécoise de lutte contre la pollution atmosphérique, STOP (Québec), or the Centre Québécois du droit de l'environnement.

Key Issues

The leading environmental issues in each province or territory reflect specific geographies and political economies. A key issue on the west coast, for example, concerns multiple threats to wild salmon stocks (from fish farms, over-fishing, forestry practices affecting spawning streams, or oil spills). Because of its northern Pacific climate, British Columbia is also home to some of the world's remaining temperate rainforests, whose preservation has been a long-standing aim of the province's environmental movement and its First Nations. The 1990s was a decade of large-scale mobilization in British Columbia against logging in old growth forests (such as in Clayoquot Sound, Carmanagh Valley, South Moresby, Great Bear Rainforest) and watersheds (e.g., in the Slocan Valley).[13]

Most recently, there has been a widespread mobilization in British Columbia against proposed pipelines that would carry bitumen from the Alberta bituminous sands to the west coast, where the product would then be exported on tanker ships. One of these projects is the Northern Gateway Pipeline that would carry Alberta bitumen to the port of Kitimat. Another project is the twinning of the already-existing TransMountain pipeline, carrying crude oil to a terminal in Burnaby Inlet, and greatly increasing tanker traffic and the risk of spills. More than 60 First Nations and aboriginal organizations (including the Tsleil-Waututh of Burnaby Inlet), environmentalists, landowners, the Burnaby municipal council, the City of Vancouver, and the Union of BC Municipalities are among the actors opposing one or both of these projects (which already have federal government approval) because of their serious risks to fisheries, livelihoods based on marine resources and tourism, human rights to health and environment, and endangered species.[14]

As the most populated and highly industrialized region of the country, southern Ontario has been the terrain of intense struggles around the banning or regulation of toxic chemicals (e.g., PCBs, PAHs, dioxin) and heavy metals (e.g., mercury, lead), solutions for waste management, drinking water and air quality, and urban sprawl. Leading issues in southern Ontario and Quebec in the 1980s included the health risks of the ubiquitous use of chemicals and industrial pollution in the Great Lakes Basin and along the St. Lawrence River.[15] During the 1980s important pieces of environmental legislation (or amendments to these) were being proposed in Ontario (e.g., the Municipal-Industrial Strategy for Abatement) and federally (Canadian Environmental Protection Act), and environmental organizations invested heavily in trying to strengthen these bills and to ensure their adoption.

As the province with most of Canada's nuclear power plants, nuclear safety concerns have been ongoing in Ontario. In Saskatchewan, there has been opposition to uranium mining since the 1970s. In New Brunswick and Nova Scotia there is a long history of opposition to the spraying of pesticides (DDT was used on the spruce budworm in the 1950s, followed by other pesticides).[16]

Another major concern in the 1980s was "acid rain" (sulphur dioxide deposits from industrial emissions). Acid rain was the impetus for the formation of the Association québécoise de lutte contre la pollution atmosphérique in 1982, and for the Canadian Coalition on Acid Rain, founded in 1981, with offices in Toronto and Washington, D.C. By 1990, when it was disbanded, the Coalition had grown to 58 member groups, representing over 2 million Canadians.[17]

Certain issues have mobilized ENGOs across the country and beyond Canada's borders, such as the depletion of atmospheric ozone caused by chlorofluorocarbon (CFC) emissions. Canadian environmentalists lobbied hard for the protocols that were eventually negotiated for CFCs (the 1987 Montreal Protocol). Another huge issue for environmental organizations across Canada in the 1980s and 1990s was the federal government's negotiation of "free trade" agreements—first with the United States (1988) and then with the United States and Mexico (1993). ENGOs joined labor organizations in opposing these agreements.

Many issues are common to environmental groups across the country today, given the characteristics of the country's geography that were described briefly in the first section of this chapter. The extensive boreal forest that crosses the northern portion of the country is highly contested terrain, as forestry, mining, and oil and gas industries seek to expand their activities, while environmentalists, ecotourism and recreation businesses, and First Nations attempt to limit and regulate such "development."[18] Indeed, a long list could be made of recent conflicts regarding "land use" decisions more generally. Provincial governments in Ontario, British Columbia, and Alberta have initiated public consultation processes leading up to land-use frameworks. Multiple "stakeholders" participate in such processes, including ENGOs, First Nations, and corporations and their associations. With the exception of the Commission on Resources and Environment in the province of British Columbia (1992–1996), these processes typically privilege "development" over ecological criteria or the rights asserted by First Nations.[19]

Regulation of agricultural and livestock-raising practices, approval of genetically modified crops, and support for organic agriculture are also issues across the country. There have been significant conflicts around intensive livestock operations (the siting, scale, and ethics of these), particularly in Quebec, where there is an influential pork producers' lobby.[20] These issues involve not only ENGOs and local citizens' groups, but also farmers' associations such as the National Farmers Union and the Union des producteurs agricoles (Quebec).

Across Canada there are also many environmental campaigns to preserve wilderness areas from industrial activities, to create provincial or national parks, and to protect the habitats of rare, threatened, or endangered species. Numerous environmental organizations focus on the monitoring and protection of waterways from industrial and agricultural sources of pollution, as well as from overuse (affecting flow rates and volumes, and hence entire aquatic ecosystems). A number of "water-keeper" coalitions have formed in recent years, inspired by indigenous values of stewardship.[21]

Campaigns for sustainable urban development link urban sprawl to greenhouse gas emissions, threats to watersheds, loss of agricultural land that could support local food, destruction of habitat, and other environmental harms. Partly because of persistent public education and lobbying work on the part of local environmental groups, a growing number of cities are initiating long-term sustainability plans.[22]

Quebec has developed a unique political culture characterized by a strong social-democratic consensus.[23] Influenced by political ecology movements in Europe, the environmental movement in Quebec—while very diverse—has manifested, especially in the last decade, a capacity to advance ecological principles as an alternative model of societal development. This convergence of ecological and social priorities has emerged out of widespread frustration with the lack of action at the federal level on climate change, opposition to large-scale commercial agriculture and its effects on landscapes, environment, human health, and quality of life, concerns about the pollution of the St. Lawrence and other waterways and lakes, and—most recently—a fierce citizen mobilization against exploitation of nonconventional oil and gas (le Mouvement Vigilance Énérgie).[24]

Climate change has become a key focus of environmental activism across the country, and from north to south. There are few issues that can be characterized as simply "local"; the preservation of boreal forest, the opposition to expansion of the Alberta bituminous sands, the demands for a transition to renewable sources of electricity in Ontario, and many other issues are being linked by environmentalists to the overarching imperative to reduce greenhouse gas emissions. In this sense, we can say that most environmental issues are "nested" within multiple spatial "scales"—from the local to the global, and thus constitute "glocal" issues.

Temporally speaking, it is evident from Tables 6.2 and 6.3 that most of the organizations that have been founded since 2002 (the year of the great "Kyoto Protocol debate" in Canada) are concerned primarily with what we might call the transition to a "post-carbon" society that reduces reliance upon fossil fuels, unnecessary consumption, and greenhouse gas emissions. Organizations that were established much earlier have also adapted their campaigns to encompass these issues.

Mapping the Canadian Environmental Movement Ideologically

Many categories are used to classify environmental organizations ideologically. Typically, the categories say something about the positions taken by organizations with regard to the dominant model of economic development and the cultural assumptions underpinning it. This model is variously described as "industrialism,"[25] "liberal-productivism,"[26] "capitalism,"[27] or the "hegemonic model of development."[28] Some organizations are highly critical of this model, while others not only take it for granted (as a quasi-natural entity), but believe that it can be reformed in various ways to make it environmentally benign. From this distinction we get such labels as "radical" versus "reformist."

Environmentalists also have a range of views about how effective liberal democracies are when it comes to preventing environmental degradation, and about how political systems may need to be reformed to incorporate ecological priorities and environmental rights. Indeed, there are multiple axes along which we might locate environmental organizations. A distinction is commonly made, for example, between "conservationist" organizations (whose focus is wilderness and wildlife preservation) and "environmental justice" organizations (which tend to focus more on the social justice [human] dimensions of environmental problems). Another significant axis positions ecocentrism at one end and anthropocentrism at the other. Ecocentrists—like deep ecologists—reject the *a priori* privileging of human interests over those of other species, and—like many indigenous cultures—emphasize the interdependence of humans and other beings.[29] Some authors classify environmental organizations by their degree of "institutionalization" as opposed to a "grassroots" orientation. Attached to these labels are often assumptions about the political or ideological orientations of these groups. For example, "institutionalized" is sometimes taken to mean "incorporated into the political mainstream," or co-opted by business and governmental interests.

Actors in environmental politics do not fit neatly into any taxonomy. If we use only one of these axes to categorize ENGOs, we are bound to greatly oversimplify what is in reality a very complex and shifting terrain of activism and coalition-building. For example, Greenpeace may be viewed as an "institutionalized" ENGO, because it has national offices, a substantial international staff, and considerable financial resources. At the same time, Greenpeace is at the "radical" end of the spectrum when it comes to its criticisms of mass-consumption capitalism and its global injustices. "Grassroots" networks of citizens' groups, on the other hand, may be focused on a single issue and may advocate fairly modest policy changes (e.g., better government monitoring of industrial polluters, the banning of lead in gasoline, or the banning of pesticides)—as opposed to the transformation of capitalism. In Canada (and across North America), there is also an "indigenous" environmental movement rooted in First Nations communities (the Indigenous Environmental Network (IEN)).

Another problem one faces when trying to classify organizations is that, although an organization may focus on one task within the broader environmental movement (e.g., conservation of wilderness), its members and staff may participate in more than one organization, and may have much broader perspectives on environmental problems than the organization's campaigns suggest. There is a certain division of labor within environmental movements, due to limited resources. It is difficult for an ENGO to take on the role of a political party (like the Green Party) that would be expected to have a comprehensive analysis of fundamental societal and environmental problems. All of these points should be kept in mind with regard to the classification of organizations within any schema.

In the Canadian case, I believe it is fair to say that most environmental activists have political affinities that would place them between social-democratic and radical post-Marxist positions on the ideological spectrum.[30] (By the latter term I mean an understanding of the ecological crisis as being rooted not only in the imperatives of capital accumulation but also in other relationships such as patriarchy, racism, and speciesism). In terms of Canada's political party system, most environmentalists would vote for either the New Democratic Party or the Green Party, when they are not voting "strategically."[31]

Most environmental advocacy and policy organizations in Canada support market-based environmental policies, but not always because they agree with market-driven politics. On the contrary, many environmentalists support policies like carbon taxes or cap-and-trade systems because they believe that these are the best policies "politically achievable." Such policies have the potential, at least, to reign in some of the environmental destruction caused by unregulated capitalist market forces. Thus, organizations like the Pembina Institute, Pollution Probe, the National Roundtable on the Environment and the Economy,[32] the International Institute for Sustainable Development, or Sustainable Prosperity call for slower rates of exploitation of resources or the use of price mechanisms to change corporate and consumer behavior. They sometimes also call for regulatory measures such as more stringent fuel emission standards or mandated quotas for renewable energies. All of these types of policies—whether termed "market-based" or "regulatory"—ultimately require governments to intervene to restructure existing markets (or even to create new markets, as with carbon offsets). Thus, most of these proposals are quite compatible with a social democratic view of the appropriate role of government vis-à-vis the economy. Some would say that these approaches fit within an "environmental management" approach that seeks to reform the dominant model of development rather than to transform it wholesale.

Conservationist organizations like CPAWS, Canadian Wildlife Federation, Canadian Nature Federation, and Union Québécois pour la Conservation de la Nature focus their efforts on legislation to protect areas considered ecologically important, to create new parks, and to strengthen species-at-risk and other biodiversity-related statutes. While they recognize the effects of industrial activities (forestry, mining) as well as climate change on the habitats of various species, they do not typically intervene directly in "climate justice" or other campaigns that demand reforms of economic regulation. Notably, the conservation organizations have come into conflict in the past with aboriginal groups over the setting aside of large tracts of land as wilderness, thereby preventing access for traditional or other uses.[33] However, there are signs that some conservation organizations are joining First Nations in joint environmental justice campaigns.[34]

Since about 2007, Canadian environmental organizations across the spectrum have devoted energy and resources to a set of closely-related issues: the expansion of mining in the Alberta oil sands, fracking for oil and gas, the building of oil and gas pipelines from Alberta to the west, east, and south, deep-water drilling for oil off all three coasts, and climate change policy. One might say that these issues have pushed environmental organizations to develop a more holistic and global analysis of local conflicts. Greenpeace, Sierra Club, Friends of the Earth, Ecojustice, Environmental Defence, the Climate Justice network, and the Indigenous Environmental Network—backed by progressive research institutes or networks such as Suzuki, Polis, Polaris, Parkland, Council of Canadians—are among the organizations that are pushing hard for alternatives to global dependence upon fossil fuels, and in the process offering critiques of the economic system and consumerist culture. The climate justice movement in general has raised awareness of the global inequities that are produced by the capitalist model of development, and that have accelerated and deepened in the wake of the liberalization of international trade and financial transactions.

Thus the agenda and the discourse of environmental actors in Canada have been significantly radicalized in recent decades by deepening connections with social justice movements in the global South. The campaigns against TransCanada's Keystone XL pipeline extension project and against Alberta's "dirty oil" exemplify the linkages being made among climate change movements internationally and between environmental and other social movements—in particular, the struggles for indigenous autonomy and cultural survival. In Canada, at least, indigenous knowledge and First Nations' struggles for self-determination (in Canada and elsewhere) are playing a much greater role in environmental campaigns than they did 20 years ago. Moreover, it is mainly via indigenous discourses that ecocentric and organicist conceptions of nature—e.g., the rights of other "nations,"[35] the rights of "Mother Earth"[36]—are reestablishing footholds in policy debates. Both of these trends are evident in the Idle No More Movement that originated in Canada in 2012 and has now gained international support and recognition. Idle No More reasserts rights established by the Treaties and imbedded in the Canadian Constitution, and seeks to use these rights to protect land, water, and aboriginal communities from the degradation caused by resource-extraction-driven development. This is a moment of unprecedented convergence between the environmental and aboriginal sovereignty movements in Canada, fuelled, in part, by the radically pro-development policies of the Conservative government that has been in office since 2006.[37]

Neither has the environmental movement in Canada been impervious to the radicalization of youth in relation to deepening social inequalities, insecurity, and ecological crisis. All of these problems are being linked—as we saw recently in the Occupy movements (beginning with Occupy Wall Street)—to the antidemocratic nature of political institutions. The newest generation of activists has made use of social networking technologies to communicate rapidly across space, and has invented creative, media-savvy modes of protest and messaging. The internet has become a critical resource for this complex web of activists, providing information and means of communication that were unavailable only 30 years ago. Online publications like *Tyee*, blogs like desmogblog.com, and websites like 350.org and oilsands watch.org are widely used by environmental activists to find coverage of events (even live streamed) and critical analyses of countless issues. We are also seeing an increased use of youth-led civil disobedience and other disruptive actions to demand governmental action on green transition.

Green Transition versus the Promethean State

Despite all this coalition-building and youth activism, some on the Left argue that the predominant orientation of the Canadian environmental movement continues to be "reformism," that is, market-based approaches to the reform of capitalism that serve to legitimize the basic logic of the existing system, and that fall far short of what is needed to achieve, for example, rapid and radical reductions in greenhouse gas emissions. Indeed, there are different policy stances within the movement with regard to the capitalist model of development. While a handful of coalitions and organizations are calling for a moratorium on oil sands expansion and massive, immediate investment in renewable energies and energy conservation measures, and might therefore be placed at the radical end of the spectrum, others adopt an insider strategy, arguing for the pricing of carbon, slowing the pace of oil sands expansion, and phased targets for transition to renewable sources of energy. However, the current Conservative government has so far been unwilling to entertain even the market-based environmental approaches, and has instead thrown its weight behind the neoliberal, promethean policies advocated by its counterpart in fossil-fuel-rich Alberta. In such a political context, it is difficult to judge whether the environmental movement is divided ideologically or is simply employing a two-pronged approach to influencing the direction of change.

It is interesting to note that the term *sustainable development*—so ubiquitous following the Brundtland Report in 1987—is now seldom used by environmental organizations. Whatever radical cachet the term had has long since worn off, which likely explains its replacement—among contemporary environmental activists—by such terms as *climate justice* or *ecological debt*—terms that embody awareness of the particular responsibility of citizens of the global North to alter their own behaviors and to reform global institutions. In terms of historical trends, we also see a movement from the conservation focus of Canada's earliest ENGOs to an environmental justice emphasis.

In recent decades the Canadian environmental movement has become much more aware of the multi-scalar nature of its struggles and of the intersecting nature of ecological and social crises. The very nature of the climate change issue—its roots in globalizing "fossil capitalism"—and the growing urgency of action to stop global warming have to some extent brought about a convergence of social movements that has altered each of them. Even the wilderness conservation organizations have begun to question—if not the entire model of development—at least the political institutions that render environmental concerns so marginal.

The current juncture of global economic and ecological crisis has created a highly unstable political moment. On the one hand, the Conservative governments in office at provincial and federal levels are pushing hard to extend and accelerate the very economic forces that have led us to this place. Environmental campaigns risk being marginalized by a governmental discourse that sees the solution to economic insecurity as being more economic growth. In the Canadian case, this economic growth is driven largely by the exploitation and export of mineral, fossil, and forestry resources. On the other hand, the credibility of governing elites has been severely undermined, globally, by the recent crises and their social consequences. Environmental organizations in Canada are also working hard to make the case that they have an alternative vision—a better future—on offer. With CO_2 levels now above 350 ppm, critical ecosystems facing collapse, and biodiversity threatened everywhere, the stakes of this struggle have never been greater.

Notes

1. The author wishes to acknowledge the research assistance of Mr. Eyene Okpanachi.

2. Department of Natural Resources, Government of Canada, The Atlas of Canada, http://atlas.nrcan.gc.ca/auth/english/learningresources/facts/supergeneral.html.

3. Ibid., http://atlas.nrcan.gc.ca/auth/english/learningresources/facts/lakes.html.

4. Ibid., http://atlas.nrcan.gc.ca/auth/english/maps/environment/forest/forestcanada.

5. McClenaghan, "Court Updates: Harmonization Accord."

6. Doyle-Bedwell and Cohen, "Aboriginal Peoples in Canada," 169–206.

7. Boyd, "No Taps, No Toilets," 1–3.

8. The Canadian Environmental Network/Reseau canadien de l'environnement uses the French acronym: RCEN.

9. Somewhat surprisingly, we see from Table 6.1 that the first provincial environmental network was established in Alberta in 1980, and that Alberta ENGOs outnumber those in the more populated province of British Columbia (which is thought to have a strong environmental movement). However, these numbers reveal only the number of affiliated organizations, and not their aggregated membership figures. That is, the number of persons claiming to belong to an environmental organization might be far higher in British Columbia than in Alberta.

The Yukon takes the cake, with by far the largest number of environmental organizations per 100,000 members of the population (37.5). At the other end of the scale, we find Quebec, closely followed by Ontario, both having fewer than one ENGO per 100,000 members of the population. Again, the figures in column 3 do not include the numerous citizens' organizations that may not be affiliated with the provincial environmental networks. For example, a coalition called Green Ontario published a report in 2003 that claimed that the coalition comprised "over 700 small and large environmental non-governmental organizations (ENGOs) including national, provincial, regional and neighbourhood-based groups" as well as "numerous other non-governmental organizations (NGOs) with an environmental mandate or interest, including cultural, health, professional and labour groups, service clubs, religious institutions, community and ratepayers associations, school clubs, and government agencies such as conservation authorities" (http://www.weconserve.ca/pdf/GOstrategy.pdf, p. 1).

10. It should be noted that not all local environmental groups are affiliated with one of the provincial or territorial environmental networks. The networks estimate that there are hundreds of local groups with whom they have informal links.

11. Some ENGOs, however, like Environmental Defence, do accept sponsorship funding from private firms.

12. This agreement was ended by Environment Canada in 2010.

13. See Magnusson and Shaw, *Political Space,* and; Wilson, *Talk and Log.*

14. Sean Kolenko, "Oil and Water: 'People of the Inlet' Oppose Kinder Morgan Pipeline Expansion," *Burnaby NewsLeader,* April 13, 2012, http://www.burnabynewsleader.com/news/147199625.html. Recent polls have found that 72 percent of residents in the Burnaby-Douglas federal riding, and 69 percent of British Columbians oppose the expansion of the Kinder Morgan pipeline. Wanda Chow, "72 percent of Burnaby-Douglas opposes Kinder Morgan pipeline expansion," *Burnaby NewsLeader,* January 19, 2012, http://www.burnabynewsleader.com/news/137683728.html. Wanda Chow, "Burnaby Council Officially Opposes Pipeline Expansion," *Burnaby NewsLeader,* May 31, 2012, http://www.burnabynewsleader

.com/news/155994595.html; Alexis Stoymenoff, "First Nations Resist Enbridge Northern Gateway Pipeline with Cross-Canada Freedom Train," *Vancouver Observer*, April 28, 2012, http://www.vancouverobserver.com/sustainability/2012/04/28/first-nations-resist-enbridge-northern-gateway-pipeline-cross-canada.

15. Adkin, *Politics of Sustainable Development*, 28–115.

16. Sandberg, *Trouble in the Woods.*

17. University of Waterloo Library, Special Collections, Canadian Coalition on Acid Rain fonds, http://www.lib.uwaterloo.ca/discipline/SpecColl/acid/ (Accessed November 6, 2011).

18. See: Pratt and Urquhart, *Last Great Forest;* Richards, Sherman, and Gismondi, *Winning Back the Words,* and Blaser, Feit, and McRae, *In the Way of Development.*

19. See: Adkin, "Democracy from the Trenches," 290–319; Fluet and Krogman "Limits of Integrated Resource Management," 123–39, and Mason, *Environmental Democracy* (for an account of the CORE experiment in British Columbia).

20. Berny, Hudon, and Ouellet, "Regulating Farm Pollution in Quebec," 33–50.

21. Five waterkeeper communities cross the boundaries of the Northwest Territories, Alberta, Saskatchewan, and Manitoba. See their website: http://www.keepersofthewater.ca/. There are also nine Canadian member organizations of the Waterkeepers Alliance founded by Robert Kennedy Jr. in the United States. See http://www.waterkeeper.org/ht/d/sp/i/238/pid/238/display/ContentDetails/i/1279/pid/238.

22. See, for example, Edmonton, *The Way We Green,* 1–82; Calgary, imagine Calgary Plan, 1–207, and Robinson, "Urban Sustainability in Canada," 159–82.

23. Quebec has had more broadly-based and militant labor and women's movements than other parts of Canada, and these have played a significant role in shaping social and family policy in the province.

24. Batellier and Sauvé, "La Mobilisation des Citoyens," 49–58.

25. Dryzek, *Politics of the Earth.*

26. Lipietz, *Towards a New Economic Order.*

27. Panitch and Leys, *Coming to Terms with Nature.*

28. Adkin, "Counter-hegemony and Environmental Politics," 135–56.

29. Eckersley, "Ecocentric Discourses," 155–86.

30. A small number of organizations (see Energy Probe Research Foundation in Table 6.2) and a growing number of think tanks (Fraser Institute, Canadian Centre of Policy Studies, Frontier Center for Public Policy, the Natural Resources Stewardship Project, Macdonald-Laurier Institute) adopt free-market liberalism positions.

31. Because Canada still has a first-past-the-post electoral system, green voters and social democratic/left voters will sometimes vote for the Liberals or the Bloc Québecois rather than for their "first choice" party. There is little published research on the green vote in Canada. See Mike Painter-Main, "Environmental Voting in Canada: Evidence from the 2006 Canadian Federal Election," paper presented to the Annual Conference of the Canadian Political Science Association,

Carleton University, Ottawa, ON, May 27–29, 2009, http://www.cpsa-acsp.ca/papers-2009/Painter-Main.pdf.

32. In May 2012 the federal Conservative government announced its intention to end its funding for NRTEE, along with its decisions to challenge the charitable status of ENGOs and to defund federally-sponsored scientific research.

33. For accounts of such conflicts, see: Ballamingie, "First Nations, ENGOs, and Ontario's Lands for Life," 84–102, and Robinson et al., "Support for First Nations' Land Claims," 579–98.

34. CPAWS, for example, has joined with Ecojustice and the Matawa and Mushkegowuk First Nations to demand the cumulative environmental impact assessment of new mining applications in northern Ontario. See: CPAWS, "Assess impacts of mega-mines in Northern Ontario, CPAWS and Ecojustice tell government," media release, November 7, 2011, http://cpaws.org/news/assess-impacts-of-mega-mines-in-northern-ontario-cpaws-and-ecojustice-tell.

35. Arquette et al., "Restoring Our Relationships for the Future," 332–50.

36. Merchant, *Death of Nature.*

37 The Idle No More Movement calls for its own chapter. When this chapter was first drafted, Idle No More had not yet been heard of; today, it has assumed leadership of the Canadian environmental movement, embraced by ENGOs. For an introduction, the movement's website is http://idlenomore.ca/.

Bibliography

Adkin, Laurie E. "Counter-Hegemony and Environmental Politics in Canada." In *Organizing Dissent: Contemporary Social Movements in Theory and Practice*, edited by William K. Carroll, 135–56. 1st ed. Toronto: Garamond Press, 1992.

Adkin, Laurie E. "Democracy from the Trenches: Environmental Conflicts and Ecological Citizenship," In *Environmental Conflict and Democracy in Canada*, edited by Laurie E. Adkin, 298–319. Vancouver: University of British Columbia Press, 2009.

Adkin, Laurie E. *The Politics of Sustainable Development: Citizens, Unions, and the Corporations.* Montreal; New York: Black Rose Books, 1998.

Arquette, Mary, Maxine Cole, and the Akwesasne Task Force on the Environment. "Restoring Our Relationships for the Future." In *In the Way of Development: Indigenous Peoples, Life Projects, and Globalization*, edited by Mario Blaser et al., 332–50. London: Zed Books, 2004.

Ballamingie, Patricia. "First Nations, ENGOs, and Ontario's Lands for Life Consultation Process." In *Environmental Conflict and Democracy in Canada*, edited by Laurie E. Adkin, 84–102. Vancouver: University of British Columbia Press, 2009.

Batellier and Sauvé. "La Mobilisation des Citoyens autour du Gaz de Schiste au Québec: Les Leçons à Tirer." *Gestion* 36 (2009/10): 49–58.

Berny, Nathalie, Raymond Hudon, and Maxime Ouellet. "Regulating Farm Pollution in Quebec: Environmentalists and the Union des producteurs agricoles

Contest the Meaning of Sustainable Development." In *Environmental Conflict and Democracy in Canada*, edited by Laurie E. Adkin, 33–50. Vancouver: University of British Columbia Press, 2009.

Blaser, Mario, Harvey A. Feit, and Glenn McRae, eds. *In the Way of Development: Indigenous Peoples, Life Projects, and Globalization.* London: Zed Books, 2004.

Boyd, David R. "No Taps, No Toilets: First Nations and the Constitutional Right to Water in Canada." Report for Ecojustice, Executive Summary, 2011, 1–3. http://www.ecojustice.ca/files/first-nations-right-to-water/view?searchterm=Boyd.

Calgary, City of. "ImagineCalgary Plan for Long Range Urban Sustainability." Calgary, Alberta: September 2007. http://www.calgary.ca/PDA/DBA/Documents/planning_policy_information/long_range_urban_sustainability_plan.pdf.

Doyle-Bedwell, Patricia, and Fay G. Cohen. "Aboriginal Peoples in Canada: Their Role in Shaping Environmental Trends in the Twenty-first Century." In *Governing the Environment: Persistent Challenges, Uncertain Innovations*, edited by Edward A. Parson, 169–206. Toronto; Buffalo; London: University of Toronto Press, 2001.

Dryzek, John S. *The Politics of the Earth*, 2nd ed. New York: Oxford University Press, 2005.

Duplessis, Josée. "Recherche sur le mouvement environnemental québécois: Rapport Final." Rapport pour le Réseau canadien de l'environnement. September 2005.

Eckersley, Robyn. 2004. "Ecocentric discourses: Problems and Prospects for Nature Advocacy." *Tamkang Review* 34, no. 3–4 (2004): 155–86.

Edmonton, City of. "The Way We Green: The City of Edmonton's Environmental Strategic Plan." Edmonton, Alberta: May 2011. http://www.edmonton.ca/city_government/city_vision_and_strategic_plan/the-way-we-green.aspx.

Fluet, Colette, and Naomi Krogman. 2009. "The limits of integrated resource management in Alberta for Aboriginal and environmental groups: The Northern East Slopes Sustainable Resource and Environmental Management Strategy." In *Environmental Conflict and Democracy in Canada*, edited by Laurie E. Adkin, 123–39. Vancouver: University of British Columbia Press, 2009.

Lipietz, Alain. *Towards a New Economic Order: Postfordism, Ecology, and Democracy.* New York: Oxford University Press, 1992.

Magnusson, Warren, and Karena Shaw, eds. *A Political Space: Reading the Global through Clayoquot Sound.* Minneapolis: University of Minnesota Press, 2003.

Mason, Michael. *Environmental Democracy.* London: Earthscan, 1999.

McClenaghan, Theresa. "Court Updates: Harmonization Accord." *Intervenor* (online newsletter of the Canadian Environmental Law Association) 25, no. 2 (June 2000). http://www.cela.ca/article/harmonization-accord/court-updates-harmonization-accord.

Merchant, Carolyn. *The Death of Nature: Women, Ecology, and the Scientific Revolution.* San Francisco: Harper & Row Pubs, 1980.

Nature Québec et STOP. "Mémoire présenté au Bureau d'audiences publiques sur l'environnement (BAPE): Développement durable de l'industrie des gaz de schiste au Québec." November 11, 2010. http://www.bape.gouv.qc.ca/sections/mandats/Gaz_de_schiste/documents/DM166.pdf.

Panitch, Leo, and Colin Leys, eds. *Coming to Terms with Nature: Socialist Register 2007.* London: The Merlin Press, 2006.

Pratt, Larry, and Ian Urquhart. *The Last Great Forest: Japanese Multinationals and Alberta's Northern Forests.* Edmonton: NeWest Press, 1994.

Richards, Mary, Joan Sherman, and Michael Gismondi. *Winning Back the Words: Confronting Experts in an Environmental Public Hearing.* Toronto: Garamond Press, 1993.

Robinson, Joanna L., D. B. Tindall, Erin Seldat, and Gabriela Pechlaner. "Support for First Nations' Land Claims amongst Members of the Wilderness Preservation Movement: The Potential for an Environmental Justice Movement in British Columbia." *Local Environment* 12 no. 6 (2007): 579–98.

Robinson, Pamela J. "Urban Sustainability in Canada: The Global-Local Connection." In *Environmental Challenges and Opportunities: Local-Global Perspectives on Canadian Issues*, edited by Christopher D. Gore and Peter J. Stoett, 159–82. Toronto: Emond Montgomery Pubs. Ltd., 2009.

Sandberg, L. Anders. *Trouble in the Woods: Forest Policy and Social Conflict in Nova Scotia and New Brunswick.* Acadiensis Press, 1992.

Wilson, Jeremy. *Talk and Log: Wilderness Politics in British Columbia.* Vancouver: University of British Columbia Press, 1998.

Chapter 7

Environmentalisms in the United States[1]

Robert Brulle

> But none of us—singly or sitting in committee—can possibly blueprint a specific "plan" for resolving the environmental crisis. . . . The world is being carried to the brink of ecological disaster . . . by the phalanx of powerful economic, political, and social forces that constitute the march of history. Anyone who proposes to cure the environmental crisis undertakes thereby to change the course of history. . . . But this is a competence reserved to history itself, for sweeping social change can be designed only in the workshop of rational, informed, collective social action. That we must act is now clear. The question which we face is how.[2]

Over 40 years ago, Barry Commoner recognized that the environmental crisis was fundamentally a social challenge and pointed to collective action by social movements as the path toward a livable planet. The attempts to deal with the problem of ecological deterioration has spawned a virtual environmental establishment, with innumerable international, national, and local government agencies, major scientific research programs, university academic programs, think tanks, foundations, and thousands of social movement organizations. Since the first Earth Day, there have been a number of actions to deal with the problem of environmental destruction. While there has been incremental progress in reducing some of the worst forms of visible pollution, these efforts pale in comparison to the changes necessary to halt this destruction. Surveying our environmental efforts, Ponting concluded that "compared with the scale of the problems, many measures are little more than cosmetic."[3]

It is clear that the efforts of the environmental movement to create an ecologically healthy and sustainable society have been inadequate. Thus the current situation requires a rethinking and reorientation of the environmental movement's activities to develop a more efficacious political practice. In this essay, I will argue that to reverse the problem of accelerating ecological degradation, an intentional program of directed social change needs to be developed and executed to increase the social learning capability of our social order. In the first section, I discuss the structure of the U.S. environmental movement through a survey of the different worldviews that define this movement. I then critique the U.S. movement, arguing that both its dominant discursive frames and organizational structure limit its capacity to effect significant social change. I conclude with a discussion of current efforts to address global climate change, and how this grassroots effort points the way toward a revitalized environmental movement.

The U.S. Environmental Movement

The U.S. environmental movement is perhaps the largest, most long-lived, and most complex social movement in the United States. There are over 6,500 national and 20,000 local environmental organizations, with an estimated 20–30 million members. It is also the longest running social movement. The first local environmental organizations were founded before the Civil War, and several still-existing national environmental organizations, such as the Sierra Club, the National Audubon Society, and American Forests, were founded in the late 19th century.

A key perspective in understanding the diversity of the environmental movement is through the use of frame analysis. Various constructions and typologies of environmental discourse abound.[4] In most instances, these characterizations of environmental discourse are constructions of the social theorist and imposed on the actually existing discourses, with little empirical evidence offered to support these typifications. Because of their self-limiting nature, these approaches are unable to grasp the historical development or diversity of discursive frames.

Using a comprehensive historical overview and textual analysis of the development of national environmental organizations based on a discursive perspective, Brulle argues that the evolution of the environmental movement in the United States has been informed since its emergence in the mid-19th century by a number of major worldviews, defined in this analysis as "discursive frames." This analysis maintains that the development of U.S. environmental movement organizations can be seen as the outcome of contingent historical events, the development of specific discourses, and the mobilization of material resources used to create these organizations, and that these factors in combination have created the numerous, partially overlapping communities that form the current environmental movements in the United States.[5] These discursive frames form the basis for the many different forms of action, organization, and objectives within the current environmental movement. Within the environmental movement, Brulle originally proposed that there were eight significant discursive frames. The number has expanded to 11 different discursive frames over time as the range of organizations examined has expanded.[6] These different discursive frames define distinct movement sectors or "wings" that diverge in terms of their definition of problems, strategies, and methods of organization. The 11 discursive frames are: Wildlife Management, Conservation, Preservation, Reform Environmentalism, Environmental Health, Deep Ecology, Environmental Justice, Ecofeminism, Ecospiritualism, Green, and Animal Rights.

Aggregate Movement Characteristics

In looking at the environmental movement across the United States, there are clear differences in the size of the different discursive communities. The largest number of organizations is found in the long-established discursive frames of reform environmentalism, preservation, and conservation. Together, these three discursive frames represent 78 percent of environmental movement organizations. All of the other discursive frames represent 5 percent or less of the total organizations. Thus despite the considerable diversity of perspectives among the U.S. environmental movements, it is dominated by these three discursive frames.

Another approach to understanding the U.S. environmental movement is to examine its historical development over time. The current environmental movement is the result of the cumulative historical development of the different discursive communities over the past century. Up until about 1950, the founding of new environmental organizations was concentrated in the discursive frames of conservation and preservation. However, the environmental movement growth was relatively static. There was a substantial increase in the numbers of organizations being founded starting in the mid-1950s up until around 1967. This was followed by explosive growth in 1968–1970 and again in 1988–1990.

The discursive frames of preservation and conservation/ wildlife management were dominant until the end of the 1930s. In the 1940s, there was a significant rise in the number of alternative discursive organizations. This was due primarily to the increase of environmental health organizations founded during World War II. Additionally, the founding of preservation organizations dramatically declined in the 1940s. However, in the 1950s, founding of preservation organizations increased rapidly, and founding of conservation/ wildlife management organizations started a long, slow decline. Additionally, founding of reform environmental organizations started a long and steady increase, which culminated in an explosive rate of growth around 1970. Additionally, as more alternative discursive frames were developed in the 1970s and 1980s, there was a slow but steady growth in these organizations in the time period from 1960 on. Thus there are unique developmental dynamics to each discursive community. It is the cumulative impact of these different developmental dynamics that has led to the highly differentiated environmental movement we encounter today.

Manifest Destiny and the Environmental Countermovement

No description of the U.S. environmental movement would be complete without addressing the unique U.S. phenomena of a strong environmental countermovement. Countermovements are "networks of individuals and organizations that share many of the same objects of concern as the social movements that they oppose. They make competing claims on the state on matters of policy and politics and vie for attention from the mass media and the broader public."[7] These types of movements arise in response to social movements for change. The change advocates, in effect, create their own opposition. By advocating an alternative discourse and challenging established interests, they create conditions for mobilization of their opposition. Countermovements originate as the change movement starts to show signs of success by influencing public policy, and threatening established interests. The elite of these interests then respond to this threat by fostering countermovement to protect their interests by opposing or containing the challenging social movements.[8] As noted by Gale, these countermovements "typically represent economic interests directly challenged by the emergent social movement."[9] The countermovement organizations that emerge take the form of elite-driven efforts to mobilize economically impacted populations, or populations that share similar interests or ideologies. The discourse of the countermovement generally takes the form of appeals for the restoration of the "established myths of society."[10]

Within U.S. environmental politics, the environmental countermovement is based on the discourse of "manifest destiny." Defined in the 19th century, manifest destiny has continued to serve as the discourse of several waves of countermovements opposed to the goals and aims of the various U.S. environmental movements. This discourse provides both an economic and moral rationale for exploiting the natural environment. Key components of this discourse are:

- Nature has no intrinsic value.
- Human welfare is based on development of the natural environment.
- The natural environment is unproductive and valueless without development.
- Human labor transforms the natural environment into useful commodities.
- There are abundant natural resources available for human use.
- Humans have a right to use the natural environment to meet their needs.

The current environmental countermovement originated with efforts initiated by neoliberals in the 1970s. They sought to weaken the substantial network of environmental regulations and oversight agencies and to blunt the environmental movement's effort to strengthen this system and make it more comprehensive. Thus anti-environmentalism has been, from the start, a keystone of neoliberal antiregulatory politics. But the perceived threats posed by climate change discourse intensified this opposition, mobilizing energy companies and other related industries and broader free-market forces. Thus the environmental countermovement has emerged as a powerful actor in U.S. environmental politics.[11]

The Marginalization of U.S. Environmentalism

Like the civil rights movement, the women's movement, the peace movement, and other movements, environmentalism was reborn in the 1960s. Building on the earlier conservation and preservation movements, the decade saw a flourishing of new ideas about environmental problems and how to address them. Intellectuals like Rachel Carson and Barry Commoner developed and promulgated a new perspective that later became known as *environmentalism.* They helped the general public understand the links between environmental degradation, ecosystem processes, and human health. Environmental organizations then repackaged these ideas in an effort to energize activists and the general public, thereby bringing about major policy changes.

The first Earth Day in April 1970 showcased an extraordinary mobilization over environmental issues and consolidated momentum that, in the few short years that followed, produced an impressive record of legislative victories. During the 1970s, major environmental organizations appeared regularly before Congressional hearings and got 20 to 30 major environmental bills passed every year with relatively limited challenge from corporations and other counter-interests. Congress passed the Clean Air Act in 1970, the Clean Water Act in 1972, and the Endangered Species Act in 1973. Legislators extended or strengthened many such landmark bills soon thereafter. By the end of the 1970s, environmental activists and legislators had created a system of federal regulatory oversight and safety agencies that included the U.S. Environmental Protection Agency (EPA), the Nuclear Regulatory Agency, and the National Oceanic and Atmospheric Administration. States also set up their own counterparts to the national laws, policies, and agencies.

These landmark laws and organizational networks significantly improved environmental quality across the nation. For example, in 1972 the Clean Water Act—passed by Congress over presidential veto in a matter of hours—required that all waters in the United States be swimmable, drinkable, and fishable by 1983, and that the discharge of pollutants into U.S. waters end by 1985. This significant, hard-hitting legislation required real changes to the standard operating procedures in nearly every American community. But this legislation, and the enforcement of it, was far from perfect. Indeed, 34 years after the Clean Water Act was passed, more than half of U.S. waters remain significantly degraded and EPA found in a 2006 study of streams that only 28 percent were in "good" condition. Also according to EPA, more than 146 million residents live in areas with unhealthy levels of air pollution.

The environmental failures and shortcomings of recent years belie what appears to be a strong and vibrant movement, at least institutionally speaking. More than 10,000 tax-exempt environmental organizations are registered with the Internal Revenue Service, and they boast a combined support base of approximately 15 percent of the U.S. population. The movement has a total annual income of more than $2.7 billion and assets of more than $5.8 billion. More than 100 new organizations are formed each year to address a wide spectrum of environmental problems.

Yet, for all this support and organizational strength, the political clout of the environmental movement appears to have eroded steadily since the early 1990s. Indeed, when it comes to activists and organizers, the current situation stands in marked contrast to the 1970s when the environmental movement displayed an extraordinary ability to mobilize support in Congress and created an impressive infrastructure of safety agencies and regulatory oversight. Despite a strong organizational base and widespread public support, most critics agree that the movement's political clout has declined over the past decade. Currently, in the face of the Republican House "Tea party" based assault on environmental laws, it is clear that the U.S. environmental movement is engaged in a desperate defensive battle.

Why is this? What do these recent political defeats say about the state of environmentalism in the United States? In 2004, marketing consultants Michael Shellenberger and Ted Nordhaus published a widely circulated article titled "The Death of Environmentalism."[12] They contended that "the environmental movement's foundational concepts, its method of framing legislative proposals, and its very institutions are outmoded!" They argue that the environmental movement needs to reframe its agenda to appeal to core progressive values and create a broader, more engaged political coalition. This approach has come under criticism in that this approach focuses on strictly rhetorical reform without addressing other aspects of environmental strategy, and that the approach advocated by them is fundamentally undemocratic.[13]

Sociological research suggests that the environmental movement's seeming lack of influence stems from three specific factors: (1) the decline of liberalism and the rise of the neoconservatives, (2) rise of professional environmental advocacy, and (3) limitations of mainstream environmental discourses.

The Decline of Liberalism and the Rise of the Neoconservatives: The mobilization and impact of social movements depends on their political opportunities, defined as "the (perceived) probability that actions will lead to success in achieving a desired outcome."[14] There are three critical processes that have an impact on the nature of the political opportunity structure: (1) the availability of supportive political allies; (2) the stance of political elites toward the social movement; and (3) countermobilization levels.[15] The environmental movement grew and experienced major legislative successes in the 1960s and 1970s in an environment with a favorable political opportunity structure, which included the dominance of the Democratic Party and the mobilization of other progressive social movements, including the antiwar, feminist, civil rights, and poverty rights movements. Organized labor and these social movements supported a national center/left political coalition that controlled national politics through the late 1970s. This created influential political allies and a responsive political elite that supported a wave of major environmental legislation.[16]

Several factors, however, undermined these political opportunities.[17] The first was economic globalization following the collapse of the Soviet Union in 1990. The state's ability to enforce environmental standards was greatly weakened, since international investors could play off competing state regimes by making large-scale investment contingent on accommodating environmental standards.[18] Additionally, globalization weakened the domestic labor movement by opening up jobs to foreign laborers. This led to a decrease of political support for progressive government actions. Secondly, the business community mobilized a "right turn" by forming new policy planning groups and think tanks.[19] Allied with the business "right turn" was a growing popular countermovement centered among the neoconservatives and the religious right. They joined forces with big business to launch a highly effective campaign against liberal environmentalism and the welfare state by promoting free market ideology. This took the form of an anti-environmental countermovement. This movement presented itself as the true environmentalists who would promote wise use of the environment in place of oppressive government regulation.[20]

This countermobilization presented two major challenges to the environmental movement. First, the rise of conservative think tanks has promoted policy analyses that seek to discredit the scientific basis of environmental diagnoses.[21] This has placed the environmental movement on the defensive, and fostered public uncertainty about the nature of environmental degradation. Second, with Republicans controlling the executive branch from 2000 to 2008, several well-known anti-environmentalists were appointed to run the government agencies charged with protecting the environment in both the Reagan and Bush II administrations.[22] Under these administrations, environmental research efforts were drastically cut back, and enforcement of existing environmental laws undermined. While a slight reprieve occurred at the beginning of the Obama administration, the resurgence of the Republicans, dominated by an extreme right-wing ideology, has placed the environmental movement on the defensive.

Rise of Professional Environmental Advocacy: A second problem has been the structures through which the environmental movement has mobilized. Although sectors of the environmental movement have relied on community organizing and grassroots participation, the overwhelming majority of environmental organizations are professional movement organizations that rely on professional staff to "speak for" rather than directly mobilize supporters.[23] In these "protest businesses,"[24] the professional staff uses direct mail and other appeals to raise funds from dispersed individuals, private foundations, and government agencies, which are then used to finance the advocacy efforts of the professional staff. Members are treated as donors rather than as activists who could become politically engaged. This encourages the fragmentation of environmental discourse as competing environmental organizations develop targeted direct mail campaigns organized around "designer issues" that frame "narrowly crafted messages delivered to narrowly targeted slices of the population."[25] Professional advocacy also fails to develop participatory skills and engagement on the part of supporters, who often become simply "checkbook" members. Direct participation, which increases understanding of environmental problems and willingness to participate in environmental action,[26] is discouraged. It also reinforces oligarchic structures in which a self-selecting leadership and the professional staff dominate decision making in the major national environmental movement organizations.[27] Overall, these processes blunt the mobilization and political influence of the environmental movement.

A key feature of this process is the channeling of the environmental movement into moderate mainstream discourses by foundation funding. In an examination of foundation grants to the environmental movement for 1970–2000, Brulle and Jenkins found that over 80 percent of foundation funding goes to the conservation land trusts and the staff-dominated advocacy and technical support organizations that promote liberal environmentalism.[28] Less than 20 percent of the funding goes to grassroots organizations, which have structures allowing for membership participation, and less than 15 percent goes to the new alternative viewpoints, such as environmental justice or deep ecology. The foundations indirectly control the environmental organizations by facilitating the growth of professional advocacy and moderate mainstream discourses, creating "a new civic America largely run by advocates and managers without members."[29] Environmental advocates are more likely to adopt "reasonable" positions and predictable behavior, which often blunts the political impact of their efforts.[30] There are, of course, grassroots and direct action sectors of the environmental movement. However, these more democratic alternative groups constitute less than 10 percent of existing environmental movement organizations, which limits their political influence.[31]

Limitations of Mainstream Environmental Discourses: The third factor contributing to the political marginalization of the environmental movement are the limitations of the dominant worldviews of the environmental movement itself. The discursive framing perspective argues that language and cultural beliefs are critical to the effectiveness of social movements.[32] Research has shown that for the environmental movement, ideologies and frames are often more critical to the political effectiveness of environmental groups than their resources or political alliances.[33] In fact, frames structure and make meaningful the strategic decisions that movement leaders make about the nature of the problems they address and how they go about this. As discussed earlier, the U.S. environmental movement is organized around three dominant discursive frames: (1) conservation; (2) preservation; and (3) reform environmentalism. A closer look at these discursive frames shows their limitations in animating meaningful action to counter environmental degradation.

Conservation started with George Perkins Marsh and Giffort Pinchot in the late 19th century.[34] Challenging "manifest destiny" assumptions, this discourse conceptualized nature as a resource to be used by human society to meet human needs. The aim is to use nature to achieve the greatest good for the greatest number of humans for the longest period of time. Government action is necessary, since corporations, driven by market considerations, will systematically exploit and degrade the environment. Thus the state is the arbiter of the public good by assuming overall scientific management and safeguarding of natural resources.[35] This discourse was part of the larger progressive movement, which promoted governmental regulation of markets to provide for social justice and the common good. Government regulation was one part of this effort to ensure that natural resources were used to advance the interests of the nation's population as a whole, and not for the economic advantage of private corporations. At its core, conservation maintains that economic growth and environmental sustainability can be reconciled by the application of scientific management.

This approach has several limitations. First, over the past 100 years, conservation has lost its connection to the larger progressive political tradition from which it originated. Lacking this connection to manage natural resources for the overall common good, conservation has become a method for scientific management of natural resources in service of the existing social structure. Instead of challenging the destructive tendencies of the existing order, this approach lacks a vision of an alternative order that could be ecologically sustainable.[36] Second, it is anthropocentric. The natural world is valued only by humans. Other beings are not worthy of consideration in their own right. Thus it advances a strictly utilitarian calculus for biodiversity.[37] Finally, it has failed the test of time.[38] Conservation has been a dominant environmental discourse for over one hundred years. Yet during the period, environmental change and ecological degradation have accelerated. As Everden argues: "having tried the Pinchot experiment for the better part of a century, and having during that period sunk deeper and deeper into environmental despair, we still are not ready to concede that perhaps the experiment has failed. How much longer must it be tested before we conclude that it is not enough?"[39]

Preservation focuses on the protection of wilderness and biodiversity. Henry David Thoreau is commonly cited as the first American voice for preservation, which was later revised by John Muir and Aldo Leopold.[40] It defines both a moral and an aesthetic relationship between humans and the environment. In this discursive frame, nature has intrinsic worth in itself. It also serves as a source of human self-renewal, as it provides an extra-human vantage point for developing a vision of a just and sustainable society.[41] For Thoreau, one would go to nature to achieve a proper perspective regarding the human enterprise. From his experience in nature, Thoreau developed a powerful political message involving a radical critique of industrial civilization and the development of an alternative vision of the good life living in harmony with nature. This was intended by Thoreau to serve as a guide to the reform of existing social order. Thus preservation was founded as a vision of humans existing in harmony both with each other and with the natural world.

This discourse has lost Thoreau's original political insight as a guide to the reform of existing society and thus has only a limited impact.[42] Muir, Leopold, and contemporary preservationists have translated this discourse into an exclusive concern for the protection of biodiversity and wilderness without any connection to larger social concerns.[43] This limits the range of issues addressed and abandons preservation's utility as an alternative cultural model that could inform the creation of a just and ecologically sustainable society.[44] Without a connection to a broader social justice vision, it appeals to the recreational needs and aesthetic values of middle- and upper-class whites. This frame also fails to address the continued consumption of natural resources. While it seeks to insulate certain ecologically valuable landscapes from economic exploitation, it does not address the underlying dynamics of the market. So protection of one area without shifting the nature of consumption merely moves the problem from one location to another. Finally, the preservationist discourse, with its exclusive natural science focus, is unable to connect the loss of biodiversity to large-scale social dynamics.

Reform environmentalism originated in the late 1950s and early 1960s. This discourse is based on the insight that humanity is part of the earth's ecosystems, and thus human health is linked to the condition of the natural environment.[45] Like conservation, it sees the need for government action to protect nature from corporate use.[46] However, it replaces conservation's limited utilitarian view of nature with the idea that nature has an intrinsic value of its own.[47] Echoing the ideas of Cars and Commoner,[48] reform environmentalism initially advanced an environmental program combining democratic practice and social justice concerns with protecting the environment.[49] In the early 1970s, this progressive political tradition was gradually replaced by a technicist discourse.[50] Environmentalism came to be identified with the scientific analysis of specific environmental problems to be addressed by legal reforms and economic incentives.[51] Scientific experts became central to the "scientific management of the environment,"[52] mediating between science and politics, advising and convincing the public and decision makers in government and industry of the need for remedial action. Underlying this approach was the politically naïve belief that giving the public and decision makers adequate information on environmental issues, either through environmental education, white papers, press releases, or public addresses, would lead to change in public opinion and demands for action to address environmental issues.[53]

In the early 1980s, a number of leading environmental organizations modified this discourse by adopting the idea that market mechanisms could bring about environmental progress.[54] Called "liberal environmentalism," the proponents of this approach argued that traditional government regulations should be replaced with market mechanisms, such as emissions trading.[55] This appealed to many mainstream environmentalists and fostered a coalition with corporate interests, leading to the U.S. proposal that led to the Kyoto Treaty.

The movement to technical and market-based analyses as the core of reform environmentalism gutted whatever progressive vision reform environmentalism held.[56] Rather than engaging the broader public, reform environmentalism focuses debate among experts in the scientific, legal, and economic communities. It may provide technical solutions to specific problems but it neglects the larger social dynamics that underlie environmental degradation.

Second, this discourse has an unrealistic conception of information in the decision making process. It assumes that neutral facts and scientific evidence will carry the debate without taking into account the role of political power, assuming that education alone can automatically outweigh vested interests.[57] It is also based on a naïve and unrealistic conception of public opinion. Instead of appealing to potential constituents in a manner that resonates with their normative expectations and everyday lived experience, which are the requirements for the acceptance of a discursive frame, it offers technical scientific discourse. The result is that "modern environmentalism—is politically naive and perhaps irrelevant."[58]

Additionally, reform environmentalism fosters an elitist practice that limits public participation. Framed by technical narratives, the social and normative issues involved in the creation and solution of environmental degradation are subordinated to technical discussions.[59] Hence both citizens and decision makers play only bit parts, heeding the advice of the scientist. By creating a technocratic value-neutral discourse, it removes moral considerations from public policy, and limits public input. This "scientization of politics"[60] serves to delegitimize the voices of those who do not speak these specialized languages. The result is an "authoritarian management of the environment and society,"[61] which also limits the movements' mobilization capacity.

Finally, in the form of "liberal environmentalism" its advocates base their arguments on a series of unfounded assumptions regarding the dynamics of market competition. Specifically, it assumes that government actions in structuring market activities can delink economic growth and environmental degradation through advances in technology.[62] The failure to realize environmental improvement through technological efficiency has been well documented in a detailed study of the U.S. economy over the time period from 1905–1995.[63] Based on an analysis of the material flows into the U.S. economic system, the researchers demonstrate that "there is little evidence of per capita dematerialization of the U.S. economy. On the contrary, increased demand seems to overcompensate for efficiency gains in every case we have investigated."[64]

Blühdorn argues that liberal environmentalism is essentially a discourse to ensure economic growth and to co-opt industrialism's environmental critics. Ecological thinking, as developed by Carson and Commoner, had emerged as the most important challenge to economic rationality. However, ecological modernization has been increasingly successful in "repackaging ecological issues as economic, technical and managerial issues, thereby overcoming the assumed incompatibility between ecological and economic thinking."[65] Rather than adjusting the market to its ecological limitations, liberal environmental discourse adapts its political program to fit within market limitations.[66]

Thus all three mainstream environmental discourses have major limitations. As they have developed, they have moved toward technocratic forms of discourse and lost their ability to mobilize, support, and guide effective action.[67] Taylor argues that "the story of American environmental theory in the twentieth century is primarily the story of an increasingly obscured political vision."[68] With a few notable exceptions, most analyses of environmental degradation do not address its social, political, or cultural roots. This limitation is due to the intellectual basis of these discursive frames on the writings of natural and physical scientists. While the scientists may have great competence in their specific areas of expertise, their social and political thinking "is marred by blindness and naiveté."[69]

Thus the problem with these discursive frames is not that they are scientifically inaccurate, but rather that they are partial. While the scientific analysis of environmental problems has developed a strong critique of the ecological impact of our current social order, this alone is not sufficient to develop an alternative vision on which an ecologically sustainable society can be based.[70] These perspectives obscure the social and cultural origins of ecological degradation, and limit the range of options considered. As a result, the proposed paths to an ecologically sustainable society that then emerge are badly flawed and lack a broader vision and a plausible theory of transition to an ecologically sustainable society.[71] The end result is that "contemporary ecology movements have yet to find the public idiom and institutional format in which that argument, in the face of structural resistance and public apathy, becomes utterly compelling."[72] Thus the environmental movement "has failed to reach its central aim of changing the most fundamental principles of the capitalist growth economy and the industrial consumer society. Rather than bringing about the envisaged revolutionary change in society, it has itself undergone fundamental change, and adapted in its structures, strategies, and demands to a new Zeitgeist and the principles of late modern society."[73] In this sense, mainstream environmental discourses are systematic self-deceptions regarding both the causes of and cures for environmental degradation.[74] Their promotion only strengthens the piecemeal nature of environmental discourse and diverts our attention from the social origins of environmental degradation. Thus most environmental strategies focus on curing symptoms, not on the underlying social changes needed to bring about an ecologically sustainable society.

Toward a Revitalized Environmental Movement

Since environmental problems originate in the social order, their resolution requires social change. Yet, while there are numerous calls for our society to change, the political means of how this process could be advanced remains vague.[75] In essence, there is no plausible theory of transition to an ecologically sustainable society. Accordingly, a political theory of transition must be developed that can generate alternatives and enable a politics to satisfactorily address the problem of ecological degradation.

The sociology of social change has long studied the mechanisms by which large-scale social change has occurred. This scholarship has identified the institutions based in civil society, outside of the dynamics of the market and the state, as a key site for the origination of social change.[76] Civil society is constituted by voluntary institutions that exist outside of the direct control of both the market and the state, and is comprised of a number of different organizational forms, most notably social movement organizations. Because these organizations are based in the deliberations of their members, the institutional dynamics defined by market forces and the political power of the state are minimized in their operation. This independence forms the key to the capacity of civil society to serve as a site for the generation of democratic action.[77] Social movement organizations develop and advocate alternative discourses and practices, and present these social problems and their solutions in the public arena.[78] This puts social movement organizations at the center of the renewal and transformation of social institutions.[79] Thus institutions based in civil society are seen as a key site for a restoration of democratic control over the political and economic systems.

Social movements have historically exerted influence over collective decisions through democratic debate in the public sphere.[80] The role of the public sphere is to serve as an arena that can "subject persons or affairs to public reason, and to make political decisions subject to appeal before the court of public opinion."[81] The public sphere is constituted by open forums, such as university conferences, newspaper editorials, or internet bloggers. In these public forums, debates and discussions can identify problems, develop possible solutions, and create sufficient political pressure to have them addressed by constitutional governments. This enables the realization of social learning based on democratic deliberation.[82]

The capability of a society to learn and respond to changed conditions is thus dependent upon the generation of alternative worldviews, the open communication of these realities into the general stock of cultural knowledge, and the use of this knowledge in the development of our social institutions. To address ecological problems in our social order requires the development of open communication from a robust civil society in which environmental issues are accurately perceived, these concerns are then effectively presented by representative institutions, and it is ensured that these concerns are satisfactorily addressed by the political system. Thus a key component to the development of an ecologically sustainable society is a "radical and broadly effective democratization."[83] Environmental movement organizations are key actors in institutionalizing this process of social change. By mobilizing citizens and providing a competent, legitimate, and authentic representation of their needs, environmental movement organizations can be catalysts for the development of change to bring about a democratic and ecologically sustainable society.

Social change also involves struggles in both the political and cultural arenas. Since social order is based upon the acceptance of a common worldview, changes in social structures are facilitated through a redefinition of what constitutes the taken-for-granted common sense of society. Accordingly, the maintenance and change of cultural systems is the locus of political struggle.[84] This allows us to see the symbolic dynamics of the political community as being based on the interaction between the dominant worldview and challenges by alternatives. The ruling class's exercise of intellectual, moral, and political leadership creates a position of cultural hegemony for this class, and ensures the reproduction of its political and economic power.[85] The process of social change involves conflicts over which ideas are legitimate. This struggle takes the form of a "war of position," where the different alliances face off against each other in elections and in other cultural and political confrontations.[86] Thus social change involves both the creation and advocacy of alternative worldviews.[87]

What does this analysis mean for the environmental movement? There are three areas that this analysis points to. To meaningfully address the problem of global environmental degradation, the U.S. environmental movement needs to increase its ability to foster social change. This includes (1) developing a coherent and shared vision of an ecologically sustainable society, (2) engaging in a struggle for cultural hegemony, and (3) reconnecting with its membership.

Developing a coherent and shared vision of an ecologically sustainable society: There is a critical need for the environmental movement to reformulate itself as a substantive challenge to the cultural hegemony of the neoconservatives. The dominant discursive frames that have defined the environmental movement are seriously flawed, and unable to connect to a larger political vision.[88] So there is a need to substantially rethink and develop a cogent discursive frame for the environmental movement. It is well beyond one person's capacity to articulate this frame. However, it is clear that a revitalized environmental frame must address two key areas.

First, this frame needs to move beyond the traditional environmental frames, and deal with the causes of environmental degradation, and not just the symptoms. This shift in emphasis means a movement from "scientifically inspired policy of revealing horror scenarios to a social-science based redirection of accountability."[89] Holding social institutions accountable for the environmental consequences of their actions involves a change of the social conditions in which decisions about the natural environment are made, including notions of private property rights.

Secondly, this frame needs to acknowledge and overcome the limited mobilization capacity of the traditional environmental frames. As noted by Rochon, "Environmental organizations, for example, are heavily centered around neighborhood problems, such as the location of a polluting industry. Environmental problems that do not have site-specific effects, such as global warming and the loss of biodiversity, are far less likely to be the subject of widespread mobilization." Part of the difficulty in mobilization lies in the narrowness of environmental agendas and the lack of a larger political vision. Without a natural constituency that has a vested interest in environmental protection, environmentalism needs to appeal to the general interest. Since "environmental issues are ubiquitous in the sense that there is scarcely a social relationship that does not involve some implication for resource use, pollution, ecosystem processes, or the biosphere, then environmentalism needs to see itself as a social movement with a broad view and a broad agenda."[90] The restoration of political vision to the environmental movement thus entails a reconnection to a larger vision of a just and sustainable society and a return to the environmentalism expounded by Thoreau, Carson, and Commoner.

At its core, the restoration of a political vision to the environmental movement involves the revitalization of the enlightenment project. The enlightenment project has as its central objective the self-determination of human communities based on reasoned public deliberations. It is clear that to address environmental issues, we need to be able to have a broad-based democratic discussion to establish universal and common goals, including environmental protection.[91] Thus the enlightenment project today consists in developing the capacity to exercise effective guidance of our economic and administrative systems by democratic action. Bronner argues that this entails "fostering the will to know and fight against prejudice, the insistence upon tolerance and reciprocity, the demand for a democratic public sphere, and the accountability of institutions."[92]

Engaging in the struggle for cultural hegemony: The development of an alternative and expanded environmental vision is inadequate to foster social change on its own. It needs to be promulgated throughout society. Yet, the media channels in U.S. society are dominated by financial interests and result in a systematic distortion of the voices that are heard regarding the environment.[93] If the environmental movement hopes to compete with this phenomenon, it must create its own similar cultural institutions.[94] The naive belief that environmental education or "getting the facts out" will successfully challenge the cultural hegemony of the neoconservatives needs to be abandoned. Rather, there is a need to develop and maintain counter-hegemonic institutions, whose role will be to elaborate and defend a restored progressive environmental vision. This includes a variety of institutions, such as think tanks, media outlets, and other innovative techniques to enhance and expand the public sphere, and engage in a public dialogue to promulgate this viewpoint.

Reconnecting with its membership: Creating an ecologically sustainable society is dependent on a strong and viable democratic public sphere. However, broad-based civic participation and democratic governance cannot be brought about by expert advocacy. While it is important that expertise be applied where necessary to deal with complex situations, the use of expert knowledge needs to be contextualized as part of a community's self-understanding. The expert's role as an omnipotent seer needs to be replaced by that as a contributor to a larger body of community-based knowledge used for self-determination in the form of "democratic science," that is, a science "that is cognitively accessible and politically accountable to nonprofessional publics."[95]

How environmental movement organizations are governed thus has important consequences for the viability of this movement. First, to enhance and strengthen the democratic nature of the civic life of the communities in which these organizations work, it is important that they have open and participatory structures. A participatory structure enables the community to develop its indigenous leadership capabilities, and increases its ability to meaningfully participate in its own governance.[96] Secondly, the political capacity of social movement organization is linked to the organization's ability to mobilize its constituency. As Skocpol[97] demonstrates, organizations with open and participatory structures are better able to develop committed members, which in turn enhances the mobilization capacity, and thus the political power of the group. Thus both the mobilization capacity of an organization and the political power of a community are heavily influenced by the structural characteristics of civic associations.

The final area that is a key to the revitalization of the environmental movement is the reform of the practices of foundations to limit their control over environmental organizations. Since the long-term vitality of the environmental movement depends on popular mobilization, there is a need to rethink environmental funding so that it is more membership-directed, democratic, and responsive to the general concerns of the organizations' members. Thus the role of external funding of movement organizations also needs to be significantly reformed so that it can enhance the ability of citizens to organize and participate in their own governance.

The movement's other, crucial ailment is a failure to translate general public support and organizational strength into specific effective actions. Though not dead, the environmental movement's organizations, ideals, and projects have failed to speak to or match current environmental challenges, legislative priorities, and public policy realities. The task for the environmental movement, therefore, is not just to be a cheerleader for the grassroots, rank-and-file membership, but also—and perhaps more importantly—to devise initiatives and proposals that target specific environmental problems and actors, and then challenge and encourage its supporters to undertake them.

Successful campaigns in the movement's earlier years were organized around the workplace, schools, and churches and largely by volunteer activists but focused upon specific, meaningful issues and targets. Today's movement has some of these types of networks (the Sierra Club and local environmental justice groups, for instance) but lacks the projects and initiatives connected to people's lives and the relevance in the political culture required to mobilize real action and change. Engaging citizens in the contemporary environmental movement in the United States will require instituting local democracy and fostering civic engagement, broadening commitments and agendas, and linking environmentalism to social justice, workplace equity, and broader social protections. It ultimately could require restructuring of civic politics in America—focusing not only on passion and mobilization but also on law and public policy. But in the final, sociological analysis, getting Americans involved in a movement that will affect real environmental change will require reorganizing the environmental movement, shifting from a "top-down" structure to a grassroots approach emphasizing concrete social problems and real-world, public policy solutions.

Global Warming and the Revitalization of the U.S. Environmental Movement

The environmental movement has recently attempted to mobilize around the issue of global warming. The mass media has played a leading role. In May 2006 Al Gore's *An Inconvenient Truth* gained a wide audience and placed global warming on the national political agenda. As attention to global warming has increased in the national media, there have been a number of new environmental advocacy groups formed around this issue, such as 350.org. There are also several more conventional groups, such as the U.S. Climate Emergency Council, the Climate Crisis Coalition, and the U.S. Climate Action Network. Finally, there has been the creation of more radical environmental organizations, such as Rising Tide and Tcktcktck. The grassroots–based movement to pressure the U.S. government to address global warming has recently grown significantly. In August 2011, a series of protests and nonviolent civil disobedience successfully transformed the routine approval of a pipeline to carry tar sands into the United States from Canada into a major political issue. Additionally, an alliance between the "Occupy Wall Street" movement and the grassroots climate movement is emerging. What will emerge from this ongoing movement building is not clear. We very well may be in a situation analogous to the period from 1963 to 1967—Rachel Carson's *Silent Spring* had been published and there was a great deal of media interest in her book, but it was only beginning to translate into the increased environmental mobilization of the late 1960s and early 1970s.

Conclusion

The U.S. environmental movement is not a monolithic structure. Rather, it is composed of a number of different discursive communities. Each of these communities has its own specific issue focus. Additionally, they developed under different historical and political circumstances and have widely varying levels of economic resources. One certainty is that this is not a static movement. Over the past century, it has developed several new discursive frames and will most likely continue to do so in the future.

Organizationally and in terms of broad public support, the environmental movement has been a remarkable success. But in terms of political clout for tackling the big issues, the movement is weak, losing critical policy battles while failing to provide strategies that can transform potential public support into environmental action. This makes it all the more incumbent on the environmental movement to help the general public connect the dots between their personal, local concerns and the dramatic, global threats to which environmental groups devote their resources and energy.

The United States needs a movement that leads away from the path of continued degradation and toward ecological sustainability. But without a paradigm shift from the top-down approach where members of environmental organizations are treated as budget funders to a grassroots focus that will engage citizens to take specific actions that stem the tide of environmental degradation, our environmental movement won't have the right approach to get us there. To move forward toward the creation of a sustainable and just society, the environmental movement needs to reorient itself, and then take action to increase the social learning capability of our social order. The end result of this process would seek the creation of an effective public sphere, which would enable the possibility of debating and making democratic decisions about how we should act to create an ecologically sustainable society.[98] This re-democratization of civil society is a necessary first step toward regaining direction over our social institutions and addressing environmental degradation.[99] This is not a utopian dream. As Dr. Commoner has noted: "Since the environmental crisis is the result of the social mismanagement of the world's resources, then it can be resolved and man can survive in a human condition when the social organization of man is brought into harmony with the ecosphere."[100] This is the challenge of the 21st century.

Notes

1. This chapter is based on a paper presented at the International Sociological Association Conference (RC 24)—Double Standards and Simulation: Symbolism, Rhetoric and Irony in Eco-Politics: International Conference at the University of Bath, UK, September 2005 under the title "The U.S. Environmental Movement: Crisis or Transition." It also incorporates extended portions of text and arguments earlier developed in several publications by the author, including Brulle and Jenkins, "Fixing the Bungled U.S. Environmental Movement," Brulle, "Politics and the Environment," Brulle and Jenkins, "Civil Society and the Environment," and Brulle, "The U.S. Environmental Movement."

2. Commoner, *The Closing Circle*, 300.

3. Ponting, *A Green History of the World*, 400.

4. Dryzek, *The Politics of the Earth*, and Alcock, "Conflicts and Coalitions within and across the ENGO Community," 66–91.

5. Brulle, *Agency Democracy, and Nature*, Brulle, "Environmental Discourse and Social Movement Organizations," Brulle and Jenkins "Spinning Our Way to Sustainability?" 1–6.

6. Brulle, Turner, Jenkins, and Carmichael, "Measuring SMO Populations," 195–211.

7. Meyer and Staggenborg, "Movements, Countermovements and the Structure of Political Opportunity," 1628–60.

8. Pichardo, "The Power Elite and Elite-Driven Countermovements," 21–49.

9. Gale, "Social Movements and the State," 207.

10. Lo, "Countermovements and Conservative Movements," 107–34.

11. Antonio and Brulle, "The Unbearable Lightness of Politics: Climate Change Denial & Political Polarization," 78.

12. Shellenberger and Nordhaus, *The Death of Environmentalism*, www.thebreakthrough.org

13. See Brulle and Jenkins, "Spinning," and Brulle, "From Environmental Campaigns to Advancing the Public Dialogue," 82–98.

14. Goldstone and Tilly, "Threat (and opportunity)," 182.

15. Tarrow, *Power in Movement*, 76–90.

16. Andrews, *Managing the Environment, Managing Ourselves.*

17. Diesing, *Science and Ideology in the Policy Sciences.*

18. Crouch, *Post-Democracy.*

19. See Ferguson and Rogers, *Right Turn*, Stefancic and Delgado, *No Mercy*, NCRP *$1 Billion for Ideas*, NCRP, *Axis of Ideology*, NCRP, *Moving a Public Policy Agenda*, NCRP, *State of Philanthropy 2004*, Covington, "Moving Public Policy to the Right."

20. Helvarg, *The War against the Greens*, Austin, "Advancing Accumulation and Managing Its Discontents," 71–105, Buell, *From Apocalypse to Way of Life.*

21. NCRP, *$1 Billion for Ideas*, 15–16, 24, Stefanicic and Delgado, *No Mercy.*

22. Shabecoff, *A Fierce Green Fire*, Covington, "Moving Public Policy to the Right."

23. Zald and McCarthy, *Social Movements in an Organizational Society.*

24. Jordan and Maloney, *The Protest Business?*

25. Skocpol, *Diminished Democracy.*

26. Shaiko, *Voices and Echoes for the Environment.*

27. Brulle, *Agency, Democracy and Nature.*

28. Brulle and Jenkins, "Foundations and the Environmental Movement."

29. Skocpol, "Advocates without Members."

30. See Manes, *Green Rage*, and Szasz, "Progress through Mischief," 524.

31. Brulle, Turner, Carmichael, and Jenkins, "Measuring SMO Populations."

32. Benford and Snow, "Framing Processes and Social Movements."

33. Dalton, *The Green Rainbow*, and Carmine and Balser, "Selecting Repertoires of Action."

34. See Brulle, *Agency, Democracy, and Nature*, Taylor, *Our Limits Transgressed*, 15–27; and Oelschlaeger, *The Idea of Wilderness*, 286–89.

35. Taylor, *Our Limits Transgressed*, 20, Brulle, *Agency, Democracy, and Nature*, 145–61.

36. Taylor, *Our Limits Transgressed*, 136.

37. Taylor, *Our Limits Transgressed*, 57.

38. Taylor, *Our Limits Transgressed*, xi.

39. Evernden, "Ecology in Conservation and Conversation," 76.

40. Brulle, *Agency, Democracy, and Nature*, 161–172, Taylor, *Our Limits Transgressed*, 4–15, Oelschlaeger, *The Idea of Wilderness*, 289–92.

41. Taylor, *Our Limits Transgressed*, 131.

42. Taylor, *Our Limits Transgressed*, 131.

43. Taylor, *Our Limits Transgressed*, 81–132.

44. Taylor, *Our Limits Transgressed*, 136.

45. Brulle, *Agency, Democracy and Nature*, 173–93.

46. Oelschlaeger, *The Idea of Wilderness*, 292–301.

47. Taylor, *Our Limits Transgressed,* 51–59.

48. Carson, *Silent Spring,* Carson, "The Pollution of Our Environment," and Commoner, *The Closing Circle.*

49. Pepper, *Modern Environmentalism,* 137.

50. Kroll, "Rachael Carson's Silent Spring."

51. Taylor, *Our Limits Transgressed,* 27–50, Evernden, "Ecology in Conservation and Conversation."

52. Taylor, *Our Limits Transgressed,* 50.

53. Diesing, *Science and Ideology in the Policy Sciences,* 243.

54. Bernstein, *The Compromise of Liberal Environmentalism.*

55. York, Rosa, and Dietz, "Footprints on the Earth," 285.

56. This argument originally appeared in the DTO Earth Blog of the *New York Times*: http://dotearth.blogs.nytimes.com/2008/08/16/plea-to-climate-lab-for-social-science/?apage=2

57. Deising, *Science and Ideology in the Policy Sciences,* 239–48, Blühdorn, *Post-ecologist Politics.*

58. Taylor, *Our Limits Transgressed,* 136.

59. Backstrand, "Precaution, Scientization or Deliberation?," 101.

60. Habermas, *Toward a Rational Society,* 68.

61. Taylor, *Our Limits Transgressed,* 27–50, 49.

62. Stern, "The Rise and Fall of the Environmental Kuznets Curve," 1426, Fischer-Kowalski and Amann "Beyond IPAT and Kuznets Curves," 28.

63. Ayers, Ayers, and Warr, "Is the U.S. Economy Dematerializing?"

64. Ayers, Ayers, and Warr, "Is the U.S. Economy Dematerializing?" 80.

65. Blühdorn, "Self-experience in the Theme-park of Radical Action?" 5

66. See Torgerson, "The Uncertain Quest for Sustainability," 15; Bernstein, *The Compromise of Liberal Environmentalism,* 178–79; Blühdorn, *Post-ecologist Politics,* 30.

67. Marx, "The Environment and the 'Two Cultures' Divide," 5.

68. Taylor, *Our Limits Transgressed,* 137.

69. Enzensberger, "A Critique of Political Ecology," 389.

70. Killingsworth and Palmer, *Ecospeak,* 19.

71. Dobson, "Critical Theory and Green Politics," 192.

72. Goldblatt, *Social Theory and the Environment,* 202.

73. Blühdorn, *Post-ecologist Politics,* 30.

74. Blühdorn, "Symbolic Politics."

75. Dobson, "Critical Theory and Green Politics," 192.

76. McAdam, McCarthy, and Zald, *Comparative Perspectives,* and Sztompka, *The Sociology of Social Change.*

77. Skocpol, *Diminished Democracy.*

78. Habermas, *Between Facts and Norms,* 381.

79. Habermas, *Between Facts and Norms,* 365, and Habermas, *Communication and the Evolution of Society,* 125.

80. Habermas, *The Structural Transformation of the Public Sphere.*

81. Habermas, *The New Conservatism,* 141.

82. Habermas, *Between Facts and Norms,* 359–60, Habermas, "Modernity: An Unfinished Project," 248–49, Calhoun, "Nationalism and Civil Society," 392–93.

83. Habermas, "A Reply."

84. Bourdieu, "The Social Space and the Genesis of Groups," 729.

85. Gramsci, *Selections from the Prison Notebooks,* Mann, "The Social Cohesion of Liberal Democracy."

86. Touraine, *The Voice and the Eye,* 25–26.

87. Rochon, *Culture Moves,* 8.

88. Killingsworth and Palmer, *Ecospeak,* 79, Evernden, "Ecology in Conservation and Conversation," 72.

89. Beck, *Ecological Enlightenment,* 16.

90. Gottlieb, *Environmentalism Unbound,* 286–87.

91. Minteer and Taylor, *Democracy and the Claims of Nature,* and Bowersox, "The Legitimacy Crisis in Environmental Ethics and Politics," 84–86.

92. Bronner, *Reclaiming the Enlightenment,* 29.

93. Habermas, "Further Reflections on the Public Sphere," 455.

94. Artz and Murphy, *Cultural Hegemony in the United States.*

95. Brown, *Toward a Democratic Science,* ix.

96. Rochon, *Culture Moves,* 137.

97. Skocpol, *Diminished Democracy.*

98. Torgerson, *The Promise of Green Politics,* 160.

99. Luke, "The Death of Environmentalism or the Advent of Public Ecology?" 316.

100. Commoner, *The Closing Circle,* 299.

Bibliography

Alcock, F. "Conflicts and Coalitions within and across the ENGO Community." *Global Environmental Politics* 8, no. 4 (2008): 66–91.

Andrews, R.N.L. *Managing the Environment, Managing Ourselves.* New Haven, CT: Yale University Press, 1999.

Antonio, Robert J., and Robert J. Brulle. "The Unbearable Lightness of Politics: Climate Change Denial and Political Polarization." *Sociological Quarterly* 52 (2011): 195–202.

Artz, Lee, and Murphy Bren Ortega. *Cultural Hegemony in the United States.* Thousand Oaks, CA: Sage, 2000.

Austin, A. "Advancing Accumulation and Managing its Discontents: The U.S. Antienvironmental Countermovement." *Sociological Spectrum* 22 (2002): 71–105.

Ayers, R.U., L. W. Ayres, and B. Warr. "Is the U.S. Economy Dematerializing? Main Indicators and Drivers." In *Economics of Industrial Ecology: Materials,*

Structural Change, and Spatial Scales, edited by C.J.M. van den Bergh and M. S. Janssen, pp. 57–93. Cambridge, MA: MIT Press, 2004.

Backstrand, Karin. "Precaution, Scientization or Deliberation? Prospects for Greening and Democratizing Science." In *Liberal Democracy and Environmentalism: The End of Environmentalism?,* edited by M. Wissenburg and Y. Levy. New York: Routledge, 2004.

Beck, Ulrich. *Ecological Enlightenment: Essays on the Politics of the Risk Society.* Atlantic Highlands, NJ: Humanities Press, 1995.

Benford, R.D., and D. A. Snow. "Framing Processes and Social Movements: An Overview and Assessment." *Annual Review of Sociology* 26 (2000): 611–39.

Bernstein, S. *The Compromise of Liberal Environmentalism.* New York: Columbia University Press, 2001.

Blühdorn, I. *Post-ecologist Politics: Social Theory and the Abdication of the Ecologist Paradigm.* New York: Routledge, 2000.

Blühdorn, I. "Self-experience in the Theme-Park of Radical Action? Social Movements and Political Articulation in the Late-modern Condition." *European Journal of Social Theory* 8/3 (2005).

Blühdorn, I. "Sustaining the Unsustainable: Symbolic Politics and the Politics of Simulation." *Environmental Politics* 16 (2007): 251–75.

Bourdieu, P. "The Social Space and the Genesis of Groups." *Theory and Society* 14, no. 6 (November 1985): 723–44.

Bowersox, J. "The Legitimacy Crisis in Environmental Ethics and Politics." In *Democracy and the Claims of Nature: Critical Perspectives for a New Century,* edited by Ben Minteer and B. P. Taylor. Lanham, MD: Rowman & Littlefield, 2002.

Bronner, Stephen Eric. *Reclaiming the Enlightenment: Toward a Politics of Radical Engagement.* New York: Columbia University Press, 2004.

Brown, Richard Harvey. *Toward a Democratic Science: Scientific Narration and Civic Communication.* New Haven, CT: Yale University Press, 1998.

Brulle, Robert J. *Agency, Democracy, and Nature.* Cambridge, MA: MIT Press, 2000.

Brulle, Robert J. "From Environmental Campaigns to Advancing the Public Dialogue: Environmental Communication for Civic Engagement." *Environmental Communication: A Journal of Nature and Culture* 4, no. 1 (March 2010): 82–98.

Brulle, R. J. "Environmental Discourse and Social Movement Organizations: A Historical and Rhetorical Perspective on the Development of U.S. Environmental Organizations." *Sociological Inquiry* 65, no. 4 (Spring 1996): 58–83.

Brulle, Robert J. "Politics and the Environment." In *The Handbook of Politics: State and Civil Society in Global Perspective,* edited by Kevin T. Leicht and J. Craig Jenkins. New York: Springer Publishers, 2010.

Brulle, Robert J. "The U.S. Environmental Movement: Crisis or Transition?" Paper presented at the International Sociological Association Conference (RC 24)—Double Standards and Simulation: Symbolism, Rhetoric and Irony in Eco-Politics: International Conference at the University of Bath, UK, September 2005.

Brulle, Robert J. "The U.S. Environmental Movement." In *20 Lessons in Environmental Sociology,* edited by K. Gould and T. Lewis, 211–27. New York: Oxford, 2008.

Brulle, Robert J., and J. C. Jenkins. "Civil Society and the Environment: Understanding the Dynamics and Impacts of the U.S. Environmental Movement." In *Good Cop/Bad Cop: Environmental NGOs and their Strategies toward Business,* edited by Thomas Lyon, 73–102. Washington, DC: RFF Press, 2010.

Brulle, Robert J., and J. C. Jenkins. "Fixing the Bungled U.S. Environmental Movement." *Contexts* 7, no. 2 (2008): 14–18.

Brulle, Robert J., and J. C. Jenkins. "Foundations and the Environmental Movement: Priorities, Strategies, and Impact." In *Foundations for Social Change: Critical Perspectives on Philanthropy and Popular Movements,* edited by Daniel Faber and Debra McCarthy. Lanham MD: Rowman & Littlefield, 2005.

Brulle, R. J., and J. C. Jenkins. "Spinning Our Way to Sustainability?" *Organization and Environment* 19 (2006): 1–6.

Brulle, Robert J., Liesel H. Turner, J. Craig Jenkins, and Jason Carmichael. "Measuring SMO Populations: A Comprehensive Census of U.S. Environmental Movement Organizations." *Mobilization* 12, no. 3 (2007): 195–211.

Buell, F. *From Apocalypse to Way of Life: Environmental Crisis in the American Century.* New York: Routledge, 2004.

Calhoun, Craig. "Nationalism and Civil Society: Democracy, Diversity and Self-determination." *International Sociology* 8, no. 4 (1993): 387–411.

Carmin, JoAnn, and Deborah Balser. "Selecting Repertoires of Action in Environmental Movement Organizations: An Interpretive Approach." *Organization and Environment* 15, no. 4 (2002): 365–88.

Carson, Rachel. "The Pollution of Our Environment." In *Lost Woods: The Discovered Writing of Rachel Carson,* edited by Linda Lear. Boston: Beacon Press, 1963.

Carson, Rachel. *Silent Spring.* New York: Houghton Mifflin, 1962.

Commoner, Barry. *The Closing Circle.* New York: Alfred A. Knopf, 1971.

Covington, Sally. "Moving Public Policy to the Right: The Strategic Philanthropy of Conservative Foundations." In *Foundations for Social Change,* edited by D. Faber and D. McCarthy. Lanham, NJ: Rowman & Littlefield, 2005.

Crouch, C. *Post-Democracy.* Malden, MA: Polity Press, 2004.

Dalton, R. *The Green Rainbow.* Minneapolis: University of Minnesota Press, 1994.

Diesing, Paul. *Science and Ideology in the Policy Sciences.* New York: Aldine, 1982.

Dobson, Andrew. "Critical Theory and Green Politics." In *The Politics of Nature: Explorations in Green Political Theory,* edited by A. Dobson and Paul Lucardie, pp. 190–206. New York: Routledge, 1993.

Dryzek, John S. *The Politics of the Earth: Environmental Discourses.* New York: Oxford University Press, 1997.

Enzensberger, H. M. "A Critique of Political Ecology." In *Political Ecology,* edited by A. Cockburn and J. Ridgeway. New York: Times Books, 1979.

Evernden, Neil. "Ecology in Conservation and Conversation." In *After Earth Day*, edited by M. Oelschlaeger. Denton: University of North Texas Press, 1992.

Ferguson, T., and J. Rogers. *Right Turn: The Decline of the Democrats and the Future of American Politics.* New York: Hill & Wang, 1986.

Fischer-Kowalski, M., and C. Amann. "Beyond IPAT and Kuznets Curves: Globalization as a Vital Factor in Analysing the Environmental Impact of Socio-Economic Metabolism." *Population and Environment* 23, no. 1 (2001): 7–47.

Gale, Richard. "Social Movements and the State: The Environmental Movement, Countermovement, and Government Agencies." *Sociological Perspectives* Vl. 29, no. 2 (1986): 202–40.

Goldblatt, David. *Social Theory and the Environment.* Cambridge, UK: Polity Press, 1996.

Goldstone, J.A., and C. Tilly. "Threat (and Opportunity)." In *Silence and Voice in the Study of Contentious Politics*, edited by R. Aminzade, J. A. Goldstone, D. McAcam, E. J. Perry, W. H. Sewell, S. Tarrow, and C. Tilly. Cambridge: Cambridge University Press, 2001.

Gottlieb, Robert. *Environmentalism Unbound: Exploring New Pathways for Change.* Cambridge, MA: MIT Press, 2001.

Gramsci, A. *Selections from the Prison Notebooks.* New York: International, 1971.

Habermas, Jürgen. *Between Facts and Norms: Contributions to a Discourse Theory of Law and Democracy.* Cambridge, MA: MIT Press, 1996.

Habermas, Jürgen. *Communication and the Evolution of Society.* Boston: Beacon Press, 1979.

Habermas, Jürgen. "Further Reflections on the Public Sphere." In *Habermas and the Public Sphere*, edited by C. Calhoun. Cambridge, MA: MIT Press, 1992.

Habermas, Jürgen. "Modernity: An Unfinished Project." In *Habermas and the Unfinished Project of Modernity*, edited by M. P. d'Entréves and S. Benhabib, pp. 38–55. Cambridge, MA: MIT Press, 1998.

Habermas, Jürgen. *The New Conservatism: Cultural Criticism and the Historians' Debate.* Cambridge, MA: MIT Press, 1989.

Habermas, Jürgen. "A Reply." In *Communicative Action*, edited by A. Honneth and H. Joas. Cambridge, MA: MIT Press, 1991.

Habermas, Jürgen. *The Structural Transformation of the Public Sphere: An Inquiry into a Category of Bourgeois Society.* Cambridge, MA: MIT Press, 1989.

Habermas, Jürgen. *Toward a Rational Society.* Boston: Beacon Press, 1970.

Helvarg, David. *The War against the Greens: The "Wise-Use" Movement, the New Right, and Anti-Environmental Violence.* San Francisco: Sierra Club, 1994.

Jordan, G., and Maloney W. *The Protest Business?: Mobilizing Campaign Groups.* Manchester, UK: Manchester University Press, 1997.

Killingsworth, M. J., and Jacqueline S. Palmer. *Ecospeak: Rhetoric and Environmental Politics in America.* Edwardsville: Southern Illinois University Press, 1992.

Kroll, Gary. "Rachel Carson's Silent Spring: A Brief History of Ecology as a Subversive Subject," available at The Online Ethics Center for Engineering and Science at Case Western University, http://onlineethics.org/moral/carson/kroll.html

Lo, Clarence Y. H. "Countermovements and Conservative Movements in the Contemporary U.S.," *Annual Review of Sociology* 8 (1982): 107–34.

Luke, T.W. "The Death of Environmentalism or the Advent of Public Ecology?" *Organization and Environment* 18, no.4 (2005): 489–94.

Manes, Christopher. *Green Rage.* Boston: Back Bay Books, 1990.

Mann, M. "The Social Cohesion of Liberal Democracy." *American Sociological Review* 35, no. 3 (1970): 432–39.

Marx, L. "The Environment and the 'Two cultures' Divide." In *Science, Technology and the Environment,* edited by J. R. Fleming et al. Akron, OH: The University of Akron Press, 1994.

McAdam, Doug, John D. McCarthy, and Mayer Zald. *Comparative Perspectives on Social Movements.* Cambridge: Cambridge University Press, 1996.

Meyer, David S, and Suzanne Staggenborg. "Movements, Countermovements, and the Structure of Political Opportunity." *American Journal of Sociology* 101, no. 6 (1996): 1682–60.

Minteer, Ben, and B. P. Taylor, eds. *Democracy and the Claims of Nature: Critical Perspectives for a New Century.* Lanham, MD: Rowman & Littlefield, 2002.

NCRP. *$1 Billion for Ideas: Conservative Think Tanks in the 1990s.* Washington, DC: National Committee for Responsive Philanthropy, 1999.

NCRP. *Axis of Ideology: Conservative Foundations and Public Policy.* Washington, DC: National Committee for Responsive Philanthropy, 2004a.

NCRP. *Moving a Public Policy Agenda: The Strategic Philanthropy of Conservative Foundations.* Washington, DC: National Committee for Responsive Philanthropy, 2004b.

NCRP. *State of Philanthropy 2004.* Washington, DC: National Committee for Responsive Philanthropy, 2004c.

Oelschlaeger, Max. *The Idea of Wilderness: From Prehistory to the Age of Ecology.* New Haven, CT: Yale University Press, 1991.

Pepper, David. *Modern Environmentalism: An Introduction.* New York: Routledge, 1996.

Pichardo, Nelson A. "The Power Elite and Elite-Driven Countermovements: The Associated Farmers of California during the 1930s." *Sociological Forum* 10, no. 1 (1995): 21–49.

Ponting, Clive. *A Green History of the World, The Environment and the Collapse of Great Civilizations.* New York: St. Martin's Press, 1992.

Rochon, Thomas R. *Culture Moves.* Princeton, NJ: Princeton University Press, 1998.

Shabecoff, Philip. *A Fierce Green Fire: The American Environmental Movement.* Washington, DC: Island Press, 1993.

Shaiko, Ronald G. *Voices and Echoes for the Environment: Public Interest Representation in the 1990s and Beyond.* New York: Columbia, 1999.

Shellenberger, M., and T. Nordhaus. *The Death of Environmentalism,* http://www.thebreakthrough.org/images/Death_of_Environmentalism.pdf

Skocpol, Theda. "Advocates without Members." In *Civic Engagement in American Democracy,* edited by Morris P. Fiorina and Theda Skocpol. Washington, DC: Brookings Institution Press, 1999.

Skocpol, Theda. *Diminished Democracy: From Membership to Management in American Civic Life*. Oklahoma: Oklahoma University Press, 2003.

Stefancic, Jean, and Richard Delgado. *No Mercy: How Conservative Think Tanks and Foundations Changed America's Social Agenda*. Philadelphia: Temple University Press, 1996.

Stern, David I. "The Rise and Fall of the Environmental Kuznets Curve," *World Development* 32, no. 8 (2004): 1419–39.

Szasz, Andrew. "Progress through Mischief: The Social Movement Alternative to Secondary Associations." *Politics and Society* 20, no. 4 (1992): 521–28.

Sztompka, Piotr. *The Sociology of Social Change*. Cambridge, MA: Blackwell, 1993.

Tarrow, Sidney. *Power in Movement* (2nd ed.). New York: Cambridge University Press, 1998.

Taylor, B. P. *Our Limits Transgressed: Environmental Political Thought in America*. Lawrence: University Press of Kansas, 1992.

Torgerson, Douglas. *The Promise of Green Politics: Environmentalism and the Public Sphere*. Durham, NC: Duke University Press, 1999.

Torgerson, Douglas. "The uncertain quest for sustainability: Public discourse and the politics of environmentalism." In *Greening Environmental Policy*, edited by F. Fischer and M. Black. New York: St. Martin's Press, 1995.

Touraine, Alain. *The Voice and the Eye*. Cambridge: Cambridge University Press, 1977.

York, R., E. Rosa, and T. Dietz. "Footprints on the Earth: The Environmental Consequences of Modernity." *American Sociological Review* 68 (2003): 279–300.

Zald, Mayer N., and John D. McCarthy. *Social Movements in an Organizational Society*. New Brunswick: Transaction Press, 1987.

PART III

Africa and the Middle East

Chapter 8

Ecological Activism in Highlighting Gendered Relations: Wangari Maathai's Green Belt Movement in Kenya

Janet Muthoni Muthuki

Introduction

Environmental issues in Kenya entail a rapidly changing environment as a result of the destruction of forests for fuel-wood, building material, and other land uses in the past 30 years. Deforestation has led to the drying up of rivers and streams, loss of biodiversity, erratic rainfall, and drought. With only 8 percent arable land and 75 percent of its workforce engaged in agriculture, Kenyan farmers face growing problems of soil erosion, deforestation, water pollution, and desertification.[1] These experiences not only create ecological stress but also create a variety of social problems that place a heavy burden on the rural Kenyan household. This burden is especially felt by women because they play a dominant role in the management and use of natural resources at the grassroots level.[2] Rural women in Kenya continue to experience scarcity in relation to fuel-wood procurement and clean water for their household use[3] and have to walk longer distances in search of these resources, something that many Western women have not had to face.[4] These women, as is the case of other women in the global South, approach environmental issues from a perspective of poverty and day-to-day domestic problems. Environmental grassroots activism has however been on the rise with women being at the forefront of social change. Local organizations and grassroots movement are seen as central to effective social change and empowerment of women. Within these organizations, Kenya's rural women get involved not only in activities to manage and expand resources but also in political and social action. It is often through women leaders and organizations that women speak out against environmental destruction and advocate for a peaceful and healthy planet.[5] Such organizations and movements focus on practical needs such as provision of water and fuel sources. They also operate in environmental, economic, and social arenas.

By prioritizing the Green Belt Movement (the GBM hereafter), an environmental grassroots campaign in Kenya founded by the late Wangari Muta Maathai, a pioneer environmental and political activist, this work attempts to expand our understanding of non-Western social movements and their approach to social change. Maathai, through the GBM, was at the forefront in advocating for ecological revolution, empowerment of women, and social change in Kenya. The GBM is amongst the prominent environmental movements in Africa and became even more so when in 2004, Maathai, a leading figure in the fight for the conservation of the environment on the African continent, became the first woman out of six Africans to win the Nobel Peace Prize.[6] She joined other African Peace Prize laureates, namely Albert Luthuli, Anwar Sadat, Desmond Tutu, Nelson Mandela, Fredrik Willem de Klerk, and Kofi Annan.[7] In the following extract from my previously published work, I advance that:

> There seems to be the sense that via recourse to earth struggles, a dual struggle of women is generated from ecological activism as is reflected in the life and work of Maathai and her GBM. Maathai's work in the GBM advances a radical, socialist politics to challenge the social structures that result in the downgrading of the earth and its resources. Her work (through her activism and writings) represents, at one level, a grassroots effort dealing with a reorganisation of political, social and economic aspects related to women in Kenya. At another level, the strategies and programmes of the GBM clearly show evidence of a type of engagement with and conceptualisation of African ecodevelopment that positions women, class and the protection of the environment as integrated elements in the struggle for sustainable development in the democratic process.[8]

This chapter therefore attempts to tease out the gendered aspect of tree planting and draws some conclusions on the possible impact of the GBM in developing a critical discourse on and thinking of environmental issues from an African perspective. The chapter is structured into four main sections. The first section provides a background of environmental issues in Kenya by describing the crisis occasioned by a depleted forest cover and highlighting the significant role played by the late Wangari Maathai and her GBM in mobilizing rural women to combat environmental degradation. In the second section, which is the theoretical framework within which this work is framed, I give an examination and critique of ecofeminism and argue for a gender perspective in examining environmental issues. Using Maathai's life and work with the GBM, I highlight how she integrates a gender perspective in her movement by giving validity to women's experiences and knowledge. Gender relations are crucial to understanding the impact on the environment because gender shapes the opportunities and constraints women and men face in securing viable livelihoods and strong communities across cultural, political, economic, and ecological settings.[9] The third section provides the context for the emergence of the GBM by exploring the effect of hegemonic experiences such as colonialism and capitalism and the resultant African states on environmental issues and gender relations in Kenya. I provide a historical account of environmental issues in various historical periods—namely the precolonial, colonial, and postcolonial eras—and how gender relations were in turn configured and reconfigured within each of these periods. In the fourth section, I highlight the programs of the GBM in the areas of women's empowerment, civic education, and advocacy, and sharing the approach with other African countries.[10] Within this section I also address the conflict between state interests and gender interests with specific reference to Maathai's role in challenging the state on the issues of gender equity and social concern. I further examine how Maathai's work is a critique of and response to industrialism (which led to colonialism), capitalism, and her contribution to sustainable development by advocating for indigenous Kenyan systems. I conclude the chapter by examining the challenges as a result of the change of leadership of the GBM with Maathai's death in 2011 and the continuity of the movement's activities.

Background on Environmental Issues in Kenya

Kenya is located in Eastern Africa and has a size of 582,644 square kilometers. Kenya's relief is varied with altitudes ranging from sea level to slightly over 5,000 meters and features including plateaus, plains, highlands, the lake basin, and the rift valley. The natural vegetation of the country includes heath and moor land, savannah grassland, scrub, semidesert, and forest. Currently the total forested area in Kenya covers about 1.7 percent with scrub and semidesert covering the largest area. This is far below the level recommended by the United Nations Environment Program (UNEP), which is about 10 percent at the very minimum.[11] The Kenyan government faces the dilemma of balancing between conservation on the one hand and meeting the demands of an increasing population and development on the other. The short-term gains of clearing of forests for timber and settling landless people work against the long-term consequences of uncontrolled deforestation. The environment is an increasingly important issue in development since those living below the poverty line directly rely on the water and land resources surrounding their communities.[12]

In northern Kenya, pastoralists have lost their herds to starvation and conflicting tensions are mounting over scarce water resources.[13] The Mau Forest in the Rift Valley, the largest indigenous forest in East Africa, has lost a quarter of its 400,000 hectares, which have been cleared to settle farmers. The effects of this have been felt by surrounding farms, the energy sector, as well as Kenya's famous national parks. The Karura and Mount Kenya forests have also been under pressure from illegal logging for years with a large part of Mount Kenya having been cleared for selective logging in 1999 and settling landless people. The capital city in Nairobi has been plagued by regular water and electricity rationing, a situation that UNEP says can be redressed by increasing forest cover. In a bid to set this situation right, there have been increased signs of citizen, state, and environmental grassroots activism. The ministry of environment has called for the planting of 7.6 billion trees by Kenyan communities over the next 20 years to increase forest cover. Other recent initiatives include a short message service campaign whereby citizens can send a message at a small charge to have a tree planted in the Mau forest and raising public awareness of the importance of trees through the educational system.

These efforts appear to be heeding Maathai's call for the protection of the environment, a cause for which she fought tirelessly for many years even when previous governments were not supportive. Her political conscience was stimulated when she realized that the government had destroyed huge parts of Kenya's forests in order to plant cash crops. Maathai[14] explains that forests provide numerous services to humanity, namely ecological balance of the earth by absorbing carbon, preventing loss of soil and subsequent desertification, and offering safeguards against flooding.[15] They also provide reservoirs for genetic resources, control rainfall patterns, and serve as catchment areas for freshwater and rivers. Forests have been a source of wealth and inspiration throughout the centuries,[16] and the decline of ecosystems worldwide has led to adverse effects such as drastic climate change. Maathai recounts that her home, which is in the Mount Kenya region, was once a place vibrant with nature's biodiversity, such as a stream, tadpoles and fertile soils ensuring a regular food supply.[17] The close proximity of the stream to her home meant that she could easily fetch water. Water and food management was designated as the woman's role and it appears to have been easily carried out due to the availability of nature's abundance. The destruction of nature's abundance led to the complication of women's lives since, given their association with food and water management, they had to walk longer distances in search of clean water and food. This destruction was occasioned by the deforestation of the forests in Mount Kenya as well as of other forests in the country.

It is against this backdrop that Maathai, also known as *mama miti*, a Swahili name meaning "the mother of trees," founded the GBM—an environmental and social justice nongovernmental organization (NGO) whose focus is environmental conservation, community development, and capacity building using the tree planting campaign as its core activity. The GBM was conceived and nurtured in the National Council of Women in Kenya (NCWK)[18] in 1977.[19] The formation was in response to needs identified by Kenyan rural women, such as the lack of firewood, clean drinking water, balanced diets, shelter, and income.[20] The movement, over a period of over 30 years, has mobilized poor women to plant an estimated 30 million trees.[21] Tree planting has served to provide fuel, food, shelter, building materials, fencing material, and income to support their children's education through the sale of timber, firewood, and fodder.[22] The tree planting activity also creates employment and improves soils and watersheds.[23] Maathai, through her GBM, was at the frontline in fighting against forest destruction in Mount Kenya. In particular, Maathai was opposed to the *shamba*[24] system, which has been advocated by politicians encouraging their constituents to cultivate in forests, thereby diminishing the forest cover.[25] In Maathai's view the practice was rife with corruption and had put species, watersheds, soil, and even local climates at risk.[26] The *shamba* system was introduced by the British and maintained by Kenya's post-independence rulers.[27]

Maathai's efforts to conserve the environment, her contribution to sustainable development, democracy, and peace[28] through the activities of the GBM caught the attention of the Norwegian Nobel committee, which decided to award her the Nobel Prize.[29] In awarding Maathai the Nobel Peace Prize, the Nobel committee[30] also placed emphasis on the importance of human rights in international politics. Peace, it seems, can only last where human rights are respected. Of even greater emphasis are the rights of women and other marginalized groups in the world today. In many societies, gender concerns are usually relegated as secondary to national concerns. This consequently has an impact on policy formulation and implementation with women being marginalized. Maathai[31] corroborates further:

> I'm quite sure that, with this kind of a prize, a lot of prejudices against women are automatically removed. I can say without exaggeration that everybody in this country, and I'm sure many people in Africa are extremely happy, and are associating themselves with the prize—both men and women. And I'm sure that, at such a time, men appreciate the role that women can play. I know that, for many men in this country, they're very proud. And they associate themselves with what the women have been doing. And this is something that I had already seen in the work that many men associate themselves. So, I think that, at a certain level, when women are dealing too with real issues, and when those issues are recognized, that there is no longer the gender bias, and that both men and the women converge in their appreciation.[32]

In this speech and indeed in her work through the GBM, Maathai articulated a gender perspective within which women's knowledge, experiences, and perceptions are given validity and foregrounded in analyzing and presenting issues. Women, particularly those living in the rural areas, play a major role in managing natural resources such as soil, water, forests, and energy.[33] Land and water resources form the basis of all farming systems, and their preservation is crucial to sustained and improved food production.[34] Women's tasks in agriculture as well as in the household make them daily managers of the environment.[35] Therefore, as the world's food producers, women have a stake in the preservation of the environment and in environmentally sustainable development.[36]

Theoretical Framework: Ecofeminism and Gender

Existing literature in the area of gender and the environment[37] and ecofeminism suggests that nations with greater gender equality are more likely to protect the environment[38] since women are more environmentally progressive. Much ecofeminist literature has cited the GBM as an example of a grassroots women and environment movement that is central to effective social change, ecological revolution, and empowerment for women. Ecofeminism, which is a coming together of the radical ecology and feminist movements, sees androcentrism (male-centeredness) as the root of the oppression of both women and the environment, and is a response to the perception that both women and the environment have been devalued in Western[39] capitalist patriarchy.[40] Western patriarchy, aimed exclusively at profits, has subjugated a sustenance economy based on creative organic nature and on indigenous knowledge possessed by women.[41] This has consequently led to the impoverishment of women and the decrease of their status relative to men.[42] Some ecofeminists advance the view that women are closer to nature and hence are more suitable to environmental management. The association between women and nature is then seen as a source of women's empowerment and environmental liberation. Consequently, strategies to emancipate both women and nature have been based on reclaiming nature and women as powerful forces.[43]

Critics of ecofeminism, on the other hand, argue that ecofeminism essentializes women in that it assumes that the association between women and nature is natural and it does not examine whether the relationship has been structurally imposed upon them. As indicated by Thomas-Slayter and Rocheleau,[44] the differentiated sociocultural construction of relationships between women and men means that the connections between them and the environment work out differently for either sex.[45] MacGregor[46] expresses a discomfort with the tendency of ecofeminist discourse to naturalize and celebrate women's caring, without examining the unequal gender relations that consign them to the nurturing roles. She advances that within Western ecofeminist discourse, environmental issues have been buttressed by the belief that conservation was a moral standpoint specific to women because of their roles as mothers and domestic managers, without consideration of other motivations outside of their designated gender roles. Further, she argues that the celebration of "earthcare" as women's unique contribution to sustainability neglects to consider the importance of politics and citizenship in women's lives.

It is instructive to note that Maathai, though a woman and a mother, was also a visionary and an environmentalist (the first woman in East and Central Africa to obtain a PhD),[47] who in her own words, "sought the cause of the problem and not just the symptoms." It was her critical thinking ability to make connections between problems such as deforestation and soil erosion and the failures of Kenya's single-party state[48] and how these issues become linked to governance, corruption, and dictatorship[49] that drew the attention of the Nobel committee. The award of the 2004 Nobel Peace Prize to Maathai[50] sparked controversy because many failed to see the connection between environmentalism and peace.[51] It seemed conflicting to critics for Maathai to be awarded the prize for fighting deforestation and advocacy for democracy and women's rights at a time when the world was experiencing war, notably the conflict in the Middle East.[52] To them politics had been overlooked in the award. Through her GBM Maathai demonstrated that environmentalism and politics, though they appear different, are in fact intimately connected since environmental concerns often stem from political issues and lead to political activism. Further, both ultimately are directed toward serving the public with a focus on alleviating conflict and creating peace.

MacGregor[53] further decries the fact that ecofeminist narratives that celebrate women's nurturing values in environmental management do not pay sufficient attention to the tension between caregiving opportunities and political engagement and the cost to women in maintaining a gender division of labor. For instance, Maathai's activism cost her at a personal level as her former husband is said to have divorced her on the grounds that she was "too educated, too strong, too successful, too stubborn and too hard to control."[54] The Kenyan society believes that the typical African woman is supposed to be dependent, submissive, and not better than her husband.[55]

A gender perspective is therefore essential to understanding the way women relate with the environment as it enables one to see from the viewpoint of the roles that society has historically assigned to men and women. Such a perspective would also require an integrative approach to gender that involves an examination of how organizations work at the community level based on gender-defined roles and relations. Environmental issues are not just the concern of women but, as Maathai[56] explains, tree planting in order to provide homesteads with wood fuel and building materials is men's responsibility as well. Consequently, policies that target women only cannot achieve the best results since the involvement of men is essential.[57] Gender relations are therefore crucial to understanding the impact on the environment. Further, gender shapes the opportunities and constraints women and men face in securing viable livelihoods and strong communities across cultural, political, economic, and ecological settings.[58] There is also a need to strengthen or establish mechanisms at the national, regional, and international levels to assess the impact of development and environmental policies on women.[59] An understanding of gender therefore enables planners to find more effective and equitable ways of managing natural resources for building productive rural livelihood systems.[60] Gender equality and equity are not only a question of fundamental human rights and social justice,[61] but are essential to the functioning of the environment as well.[62] However, in an unequal society, the impact of environmental degradation falls disproportionately on the least powerful.[63]

Gendered divisions of labor, land, and other resources have meant that women have been uniquely and disproportionately affected by ecological destruction.[64] The division of labor between men and women in the global South changed with shifts in the economic structure of production, such as the expansion of cash crops for the market at the expense of subsistence crops for the family,[65] consequently leading to the interference of the sustainable practices of African communities.[66] With the introduction of a capitalist economy, women experienced marginalization and were reduced to economic dependence.[67] Colonial rule was characterized by the exclusion of women from the newly created colonial public sphere. Colonial rule also left the legacy of a patriarchal state, a system that African male political leaders adopted at independence.[68] Consequently, African systems continue to represent systems that prioritize and uphold male privilege.[69]

An African ecological perspective therefore needs to interrogate the role of historical circumstances such as colonialism on the environment and gender roles. Thomas-Slayter and Rocheleau[70] note that the destruction of the forest cover in Kenya began with the advent of British colonial rule, which began in the late 19th century and lasted until 1963 when the country gained independence. Changes brought about by colonialism led to a shift in land uses, consequently increasing the sedimentation of major rivers, and the depletion of forests, soil and water resources available to the poor. Further to this, the colonial system interfered with the social structures of African communities, especially gender relations in agriculture and environmental management. In providing the context for the emergence of the GBM, I now turn to the role of colonialism on the environment and gender relations by examining these relations in precolonial, colonial, and postcolonial Kenya.

A Historical Account of Environmental Issues and Gender Relations in Kenya

Precolonial Kenya

The pre-colonial period in Kenya was characterized by a great amount of wildlife and plant species as well as human communities that sustained themselves and their environment for themselves and for posterity.[71] The communities and the environment did not coexist easily but the pre-colonial people purposefully made use of certain land management strategies, forms of traditional[72] knowledge and mutually beneficial relationships with various cultural groups to maintain ecological and sociocultural sustainability.[73]

Wamalwa,[74] in her examination of traditional natural resource management systems in Eastern Africa, argues that Kenyan pre-colonial society was geared toward conservation.[75] During pre-colonial Kenya, land was communally owned[76] through indigenous forms of land tenure and natural resource management systems.[77] Individuals did not have private rights to land. Instead, a political group or lineage authority was responsible for ensuring that all community members had fair access to the land.[78] The fair allocation was done through clan lineage.[79] Each clan was given a specific area and all the members, including the women who were married to members of the clan, were able to acquire land for use. This ensured that every household had access to plots in diverse micro-ecosystems in the area.[80]

Kenyan traditional societies therefore tended to reinforce holistic views of the world and encouraged techniques[81] that would lead to sustainable resource management for years prior to British colonial rule. These techniques did not require much in terms of specialization of occupation, though differentiation of occupations did exist between men and women. Mikell[82] indicates that the women were symbolically linked to the earth, fire and water, three of the four elements in traditional culture. This division relegated women to the more laborious tasks of food preparation, fuelwood gathering, and tilling the soil. Due to their role as food and firewood gatherers, women were then involved in developing strategies based on limited resources and population density that allowed for the healthy coexistence of communities.[83] Men were given the responsibility for the fourth symbolic element, namely the air, which carried speech and therefore gave them more opportunities for public and community leadership.

However, in the agricultural cycle, the sexual division of labor ensured that men took part in agricultural production.[84] Maathai[85] explains that both men and women practiced farming but they grew different crops. Women were in charge of annual crops, which they stored in granaries, while men were in charge of perennial crops like yams, cassava and bananas, which they stored on the farm. The social structure of these communities was such that it was difficult to make distinctions between domestic and public economic gender roles. It then appears that both women's and men's rights were carefully defined and mediated within social structures of subclan, clan, or ethnic community even though there were uneven relations of power between men and women. It seems, therefore, that Kenyan pre-colonial societies can be said to have maintained a system of ecological and social sustainability with both men and women participating in their different capacities. Though they both had differing roles in environmental management, they were both involved in the agricultural process.

Colonial Kenya

The indigenous systems of ecological and social sustainability were disrupted by British colonial rule. According to Mies,[86] industrial development in Europe necessitated permanent occupation of the colonies by colonial powers and the destruction of the local natural economy.[87] The old world order of cooperation between humans and nature was replaced by a mechanistic worldview of modern science, which led to the death of nature by the sanctioning of unchecked commercial and industrial expansion consequently leading to environmental destruction.[88] British colonial efforts in Kenya were based on a patriarchal and capitalist socioeconomic ideological framework that emphasized the domination and domestication of nature and the imposition of extractive and exploitative relationships that benefited British economic and technological expansion at the expense of the sustainability of human and natural resources.[89]

The concept of nature as a resource to be dominated and exploited for human beings was imposed upon the Kenyan land during the British colonial period with far-reaching consequences. Kanogo[90] points out that in this regard, the colonial period was characterized by a reorganization of traditional land tenure and agricultural systems and a massive redistribution of land for the benefit of white settlers that had devastating effects on the majority of Kenyan rural communities.[91] During the colonial rule, the British government allocated themselves land, which they referred to as the White highlands.[92] The Colonial Land Acts resulted in the repossession of communal land considered uncultivated and uninhabited by the British from Kenyans and made it government property.[93] The colonial government then introduced individual land ownership by demarcating land and issuing title deeds.[94] As a result, many people were dispossessed of their community land and were denied user rights when the European settlers legitimated their ownership through colonial legislation.[95]

Demarcation of land complicated women's lives in that they could no longer freely harvest branches from trees for firewood, as they were now privately owned under English law. Before the Land Adjudication Acts in Kenya of the 1950s, collecting fuel-wood was a free, communal act and there was little or no commercialization of fuel-wood or charcoal.[96] Further to this, the disruption of a collective land-tenure system and of fallow shifting cultivation[97] interfered with the free access to land use enjoyed by women through their lineage or marriage.[98] According to Chandler and Njoki,[99] colonialism denied women land and tree rights.

Further, the men moved in search of work in white settler plantations, leaving women behind to care for the family.[100] This migration of men in search of waged labor increasingly placed the responsibilities of local environmental management on women,[101] a task that according to Maathai[102] had traditionally been divided between men and women. Though women were involved in environmental management, they were alienated from the process of effectively managing these resources as they were not allowed to own natural resources, or land, or make decisions without male consent concerning these resources[103] under British patriarchal rule. Mies[104] reiterates that the sexism of British colonialism created an atmosphere of patriarchy that left women out of land use deliberation, even though they were in the best position to discuss the importance of land tenure and management from their individual sociocultural roles in most African societies.[105] British patriarchal colonial rule therefore served to heighten and reinforce gender divisions by preventing women from owning land and managing resources without male consent. Women were forced to deal with severely decreased land supply, use of cash crops for money that was destructive to the soil, and environmental degradation on an unprecedented scale.[106] Cash crop farming excluded women, as their own indigenous sustenance economies were destroyed.

Colonial rule therefore manipulated gender divisions in environmental management as it entrenched the division of the public and the private in economic gender roles. Moreover, women's labor in the home was unpaid and perceived to be subordinate to men's labor in the marketplace.[107] Mies[108] asserts that Western capitalist patriarchal structure owes its wealth to the unpaid labor of women.

Postcolonial Kenya

According to Emmons,[109] the most destructive aspect of colonialism was the internalization of colonial and capital-centered worldviews by the Kenyan people. The colonial administrators ensured that the indigenous governments, who took over the governing of Kenya, embraced capitalist principles of socioeconomic organization and would therefore continue supplying Britain with resources and cheap exports. This served to create internal colonies and perpetuate old colonial relationships. The end of the colonial period therefore did not lead to the end of British exploitation of

Kenyan resources, human labor, and systems of production. Consequently, Kenya has continued to relate with Britain in a way as to preserve British capitalist interests and socioeconomic dominance in the global economy.[110] Kenya's postcolonial presidents have been effective in perpetuating capital-based dependency relationships between the North and the South in the postcolonial era by instituting economic policies that favor foreign investments, foreign ownership of resources, and capitalist principles of production and resource exploitation in Kenya. These policies have led to environmental degradation and impoverishment. Etkins[111] corroborates that resource extraction does not benefit indigenous people but benefits the wealthy governments and foreign companies and the elites of the South resulting in great disparities in resource and income distribution.[112] The diversion of natural resources to market economies then generates a scarcity for ecological stability and creates new forms of poverty.[113] Consequently, policies favoring foreign ownership of resources lack sustainable solutions to the problems of the majority of the Kenyan people, especially the ones living in the rural areas.

Rural indigenous women in Kenya continue to experience scarcity in relation to fuel-wood procurement for their household use, which includes issues related to significant environmental strain, demanding workloads, health and nutrition concerns, and the overlooked specifics of knowledge and use requirements.[114] Deforestation and reforestation of monoculture[115] species for purposes of commercial production has led to the diminishing of indigenous forests. This has affected women's ability to maintain a subsistence household. Indigenous trees provide a variety of trees for food, fodder, household utensils, dyes, and medicines, whereas monoculture species do not.[116]

Demand for wood fuel energy has frequently been cited as the most common cause of deforestation in developing countries, at least as reflected in Western literature.[117] Firewood is estimated to provide approximately 75 percent of the energy used in sub-Saharan Africa.[118] With such high demand, the consumption of wood as a fuel outstrips the regeneration of forest cover.[119] Moreover, many family members in rural Kenya who cannot find gainful employment have resorted to charcoal burning. Charcoal burning has consequently contributed gravely to deforestation.[120] Fortmann[121] however argues that women do not typically fell trees for fuelwood, but men fell trees for charcoal sales.[122] Fuel-wood consumption for household use therefore plays a minor role in deforestation compared to farming, pasture and grazing land, timber, logging, housing, and charcoal production.[123] Further, Dankelman and Davidson[124] claim that forest planners have ensured that enough timber is supplied for building and industry but have until recently failed to plan for fuel-wood needs.

Reforestation programs and/or tree planting schemes often disseminated through government agricultural extension workers or rural development projects funded by nongovernmental agencies attempt to address the fuel-wood crisis, but to little avail.[125] Maathai[126] explains that government foresters insist on using technical terms in teaching the basics of tree-planting nursery management to semi-illiterate women. This then makes it more difficult for women to implement what they have learnt. Maathai[127] advances that women through the GBM have countered this by becoming innovative in using techniques unacceptable to professional foresters (usually male). Kenyan women have played a major role in tree planting, such as establishing exotic wood plantations like pine, eucalyptus, and cypress, which are the basis of the country's timber industry today. Both the colonial and the current education system have promoted exotic biological diversity of trees and crops for rapid economic returns at the expense of indigenous species. The GBM has confronted the challenge by persuading farmers to plant indigenous trees in order to conserve local biodiversity.[128]

However, in spite of their great role in tree planting, when it comes to paid employment, women are left with menial and less-well-paid jobs and are rarely involved in management, ownership of land, or professional positions. The main decision and policy-makers are men and hence matters of importance to women are given low priority. In the changes that have taken place—such as privatization of land and social, political and economic links to the West— men have been given more sharply defined preferences in the balance of rights and responsibilities.[129] Exclusion of women from the public sphere, a legacy of colonial rule,[130] results in policies in which women's knowledge and their interests are ignored in decision making and development planning.[131] For instance, as Dankelman and Davidson[132] claim, forest planners have ensured that enough timber is supplied for building and industry but have until recently failed to plan for fuel-wood needs.

Chiuri et al.[133] suggest that women have been disregarded because their beliefs in sustainable management challenge the capitalist mentality of the concept of nature as a commercial resource, which human beings can exploit to meet their needs. Furthermore, as Dankelman and Davidson[134] argue, the exclusion of women at policy-making levels is a result of conditioning such that certain types of jobs such as forestry are a male preserve.[135] The consequence is that the current environmental dilemma is based on an imbalanced reliance on masculine values and experience, and the power of patriarchy. Therefore, lasting and effective conservation and development of the environment must involve women, who are the primary environmental managers in Kenya, as well as their holistic, traditional approaches to natural resource management.[136] By investing in rural women, Maathai and the GBM have highlighted the significance of women's involvement in community organization given their roles in the agricultural processes, rural livelihood systems, and management of natural resources. Through tree planting Maathai inspired rural Kenyan women to establish public green belts in the spirit of self-reliance and empowerment.[137] The mission of the GBM has been accomplished through various strategic programs.

The Programs of the Green Belt Movement

Within the "Women for Change" program commissioned in early 2003, the GBM has assisted in facilitating the establishment of income-generating activities such as tree planting, bee keeping, and food processing to engender economic empowerment.[138] Tree planting has also been useful in providing a platform for women in leadership. According to Maathai,[139] through their involvement in the GBM, women gain some degree of power over their lives, especially their social and economic position and relevance in the family.[140] Maathai[141] indicates that, though deprived in various aspects, women have gained self-confidence through their work in the GBM. She states that the GBM has trained thousands of women in environmental conservation and management. As a result, some of the women have initiated group projects, for instance revolving funds, while others have become independent leaders and decision makers capable of addressing community-related issues.

Whilst tree planting has always been the central activity, the GBM programs have expanded to include civic education, advocacy, food security, and green belt eco-safaris.[142] In the area of civic education, the GBM has established a pilot civic education and advocacy project to raise public awareness on the need to protect the environment and be active participants in the political process by voting.[143] The GBM Learning Centre in Nairobi offers seminars on good governance, advocacy, culture, environment, and environmental justice.[144] Through its advocacy programs, the GBM has initiated advocacy activities since the late 1980s directed toward preventing forest destruction, ending poor governance, and ending human rights atrocities such as tribal clashes and corruption.[145] In 1997, the GBM established a Pan-African Green Network to share the GBM approach with other African countries through two-week training workshops.[146] The overall goal of the program has been to share the approach while raising awareness on the importance of conserving local biodiversity.

The Green Belt Movement and State Engagement

The GBM has played a pivotal role in highlighting gendered relations and challenging patriarchy within national and global ideological structures. The policies of the state that continue to impoverish women and decrease their status relative to that of men have given rise to the argument by a generation of feminist theorists that the state is both capitalist and patriarchal[147] and hence fosters unequal gender relations. Maathai was at the forefront in opposing the capitalist mindset of the Kenyan government.[148] As mentioned earlier, the Kenyan government has tended to favor capitalist principles of production and resource exploitation, which benefits a few individuals in government and foreign investors. The GBM has been involved in advocacy activities against environmental degradation. One such activity was the effort against the government's decision to construct a multimillion dollar high-rise complex in Uhuru Park in the middle of Nairobi city in 1989.[149] According to Maathai the construction of the multimillion dollar high-rise building would have obliterated much of Uhuru Park, one of Nairobi's largest green belts.[150]

The GBM not only highlights environmental issues but also represents an ecological activism within an African feminist model in the struggle to participate in Kenya's democratic structures. This is evident in the use of tree planting to motivate democratic reform. Indeed, in her Nobel Peace Prize acceptance speech, Maathai[151] confirmed that although initially the GBM's tree planting activities did not address issues of democracy and peace, it soon became apparent that responsible governance of the environment was impossible without democratic change and reform.[152] GBM's involvement in the clamor for democratic reform was most noticeable in the case where the GBM in 1992 gave support to the mothers of political prisoners who had demonstrated in Uhuru Park for the release of their sons. Their sons had been jailed for advocating for democracy, better management of resources, an end to corruption, and the introduction of a multiparty political system in order to develop democratic space. Trees of peace were planted at Nairobi's Uhuru Park, at Freedom Corner, and in many parts of the country to demand the release of prisoners of conscience and a peaceful transition to democracy.[153]

Women's determination and courage were equally visible as they launched a hunger strike and set up a camp in Uhuru Park at a corner they named the "freedom corner" to demand the release of their sons. Most of these were rural women, with some being illiterate. Four days after the women commenced their strike, the government decided to break up the camp and the women were violently beaten up. The beating continued even after the women, who were unarmed, defended themselves by stripping naked, an act that is believed in some Kenyan cultures to bring on a curse. Maathai[154] and Mikell[155] corroborate that African women use cultural practices that they feel will empower them. From this account, it is evident that the women's recourse to old customary methods of shaming men into socially appropriate behavior is a reflection of women's increased vulnerability as they protest against the oppressive policies of their governments. The women's hunger strike raised people's awareness of political prisoners and the need to stop human rights violations in democratizing societies. Most importantly, the strike confirmed women's ability to persevere in the pursuit of a strongly felt cause, despite the reality of violent opposition from the male leaders.

Kenyan male leaders for the most part exhibit resistance in responding to women's political demands as shown by the breaking up of the camp at Uhuru Park and the torturing of the women. Male politicians perceive the pressure to democratize and open up public participation to women as externally derived (in other words, a Western and foreign influence), and as one that produces female responses that do not stem from indigenous Kenyan experiences and gender roles. Indeed, African male leaders also feel the pressure to rewrite political agendas to encourage pluralism and to include the interests of women, the impoverished, and other diverse groups as derived from advanced Western states.[156] African women therefore have to strike a delicate balance between trying to achieve greater public representation for themselves and supporting the rights of African states to be autonomous decision makers.

The Green Belt Movement and Sustainable Development

Alongside using tree planting in pressing for democratic reform, Maathai and the GBM have been at the forefront[157] in resisting Western imperialism by opposing the imposition of genetically modified crops on African soil and instead advocating for indigenous Kenyan systems that lead to the protection of indigenous local biodiversity, knowledge, practices, and wisdom, which ensure sustainability. In line with this, the GBM is currently involved in educating farmers on sustainable farming methods, such as organic farming, to maintain soil fertility instead of using chemical fertilizers that inhibit the regenerative capacity of the soil.[158] Sustainable development is key to the work of the GBM. Sustainable development is defined by the International Council of Local Environmental Initiatives as development that delivers environmental, social, and economic services to all residents of a community without threatening the viability of natural-built and social systems upon which the delivery of these services depends.[159]

Sustainability for women, as advanced by Braidotti,[160] involves including rural African women's voices into the discourse without reductionism, essentialism, or paternalism in practical or theoretical terms in order to document the value of indigenous practices as gendered and ecological spaces and to reinvent sustainability.[161] This calls for an integration of gender concerns and perspectives in policies and programs for sustainable development.[162] Gender matters in the context of sustainable resources because the responsibilities of managing resources are designated according to gender. Throughout the world, especially in nonindustrialized countries, there is gender differentiation relative to the management of natural resources.[163] Jiggins[164] takes it further by positing that a sustainable world is one in which societies address issues of the intersecting inequities based on class, race, gender, ethnicity, as well as excessive consumption and expenditures for destruction.

Women's emancipation therefore needs to be connected with a need to challenge unequal gender relations and existing economic structures that define global exchanges that usually disempower the South, the poor, people of color, and women.[165] Through her GBM, Maathai has articulated this struggle by advocating changes in consumption patterns and the adoption of lifestyles that embrace values such as self-sufficiency instead of dependency, cooperation instead of competitiveness, and the respect for all creatures and creativity.[166] Maathai[167] challenges unsustainable patterns of consumption and production, particularly in industrialized countries as the major cause of the continued deterioration of the global environment.[168] The living standards and consumption patterns of the rich countries of the North continue to be high because of the colonized South and the continuation of its exploitation of cheap labor and resources.[169]

Globalization, which has led to the removal of state controls on trade and investment, has also led to increased inequalities between men and women.[170] Trade policies have had different consequences for women and men, because they differ in their access to economic resources due to their social responsibilities. As a result of the debt burden and other economic difficulties, many developing countries, Kenya included, have undertaken structural adjustment policies to deal with the debt burden and debt repayment.[171] This has led to the widespread retrenchment of nationals, particularly women who occupy low-cadre jobs. Muteshi[172] confirms that the gender dimensions of the costs of adjustment include intensification of women's domestic work and the interruption of children's education, especially for girls, as well as an increase in women's participation in informal work. Toward this end, the GBM has been involved in the Jubilee 2000 campaign that continues to advocate the cancellation of the unpayable debts of the less developed countries, many of which are in Africa, and owe to the rich in the North.[173]

Conclusion

Clearly, the strategies of the GBM call for an understanding of interconnectedness across the globe by making connections between environmentalism, gender, democratization, human rights advocacy, and sustainable development. The GBM has advocated for a holistic approach as the road to environmental and social change and it is within this framework that the movement appears to question whether women want to be integrated in societal systems that oppress both men and women.[174] Indeed, as Maathai[175] stated in her Nobel Peace Prize acceptance speech:

> Today, we are faced with a challenge that calls for a shift in our thinking, so that humanity stops threatening its life-support system. We are called to assist the earth to heal her wounds and, in the process heal our own, indeed, to embrace the whole creation in all its diversity, beauty and wonder. This will happen if we see the need to revive our sense of belonging to a larger family of life, with which we have shared our evolutionary process.

Such a shift in thinking requires courage and conviction, which Maathai demonstrated in her activism during her lifetime. Through the GBM, she demonstrated that leadership combined with knowledge and experience creates a powerful force for change.

With her death in September 2011 after a long and protracted battle with cancer, the GBM faces the challenge of carrying on with its vision without her visionary leadership. As Nepal[176] observes, leaders are central to social movements since they play a crucial role in defining goals and advancing strategies, forming coalitions, mobilizing followers, and resisting external repression. This in some cases leads to a situation where a movement is identified with a particular person thereby raising questions about who will bear the torch in the event of the retirement or passing on of the founder-leader. Nepal[177] however notes that as much as leaders form movements, leaders are also made by movements. Maathai, a pioneer academic who founded the GBM based on her critical ability to make connections between environmental conservation, gender, sustainable development, and democracy, also received much of her national and international acclaim through the GBM.

The GBM, which started as a grassroots environmental campaign for empowering rural women, has gone through various stages of growth and change in its more-than-30-years existence. From the initial stages of mobilizing rural women through tree planting to meet their basic needs for clean water, fuel, food, and income, the GBM progressed to national issues of pressing for democratic reform in advocacy activities. At a regional level the GBM has shared its approach with other African countries through the Pan-African Green Network. Through the GBM, Maathai also played a leading global role in challenging unsustainable consumption patterns of the North that involve exploiting the poor of the South and called for the cancellation of the backlog of debts of the poor countries in Africa.

At its current stage, the GBM has had to shift from the visionary leadership of Maathai who had been the face of the movement since its inception. The movement is now tasked with remaining a value-driven organization by maintaining a level of continuity of its vision, mission, and values. When accepting the Nobel Peace Prize, Maathai stated that it was not only a recognition of her efforts in the fight for the protection of the environment and social change but also of many others who had been working behind the scenes. The baton has now passed on to the leadership capacities developed over time within the GBM to keep fighting for the conservation of the environment and to hold governments accountable. The GBM has continued to engage in tree planting initiatives, campaigns to stop land grabbing in Kenya, as well as raising public awareness concerning protecting watersheds. In May 2012, the GBM joined with other partner organizations to launch the Kenyan chapter of the international campaign to stop rape and gender violence in conflict. Kenya is one of four countries worldwide (the others being The Democratic Republic of Congo, Burma, and Columbia) that have been selected for this initiative. This launch was held at the Uhuru Park's freedom corner where Maathai and the GBM once gave support to the mothers of political prisoners who had demonstrated for the release of their sons. Most of the GBM's activities continue to be made as a tribute to Wangari Maathai.

Notes

1. Foundation for Sustainable Development, "Environmental Issues in Kenya."
2. Rosa, "State of the Future 2005: Update on Women's Issues."
3. Chandler and Njoki, "Indigenous Gendered Spaces."
4. Rodda, *Women and the Environment.*
5. Dankelman and Davidson, *Women and Environment in the Third World.*
6. The Norwegian Nobel Committee, "The Nobel Peace Price 2004."
7. The Royal Forums, Address by Professor Ole Danbolt Mjøs, Chairman of the Norwegian Nobel Committee.
8. Muthuki, "Rethinking Ecofeminism."
9. Women and Population Division Sustainable Development Department FAO, "Participation and Information: The Key to Gender-responsive Agricultural Policy."
10. Maathai, *The Green Belt Movement* and Women and Population Division Sustainable Development Department FAO, "Participation and Information: The Key to Gender-responsive Agricultural Policy."
11. Gilson, "Root Causes."
12. Foundation for Sustainable Development, "Environmental Issues in Kenya."
13. Foundation for Sustainable Development, "Environmental Issues in Kenya."
14. Maathai, Nobel lecture.
15. Maathai, Nobel lecture.
16. Maathai, Nobel lecture.
17. Maathai, *The Green Belt Movement.*
18. NCWK was formed in 1964 to forge a common front through which common issues and objectives could be promoted, especially those concerning women. NCWK was a forum where African women could meet and discuss their future in postcolonial Kenya. Currently it provides a common forum for action for many of the national women's organizations and groups, both rural and urban.
19. The Ogiek people /ecoterra, "Nobel Laureate Maathai Links Environment to Peace, Democracy."
20. Muthuki, "Challenging Patriarchal Structures," 83–91.
21. The Royal Forums, Address by Professor Ole Danbolt Mjøs, Chairman of the Norwegian Nobel Committee.
22. The Ogiek people /ecoterra, "Nobel Laureate Maathai Links Environment to Peace, Democracy."
23. Maathai, *The Green Belt Movement;* The Ogiek people/ecoterra, "Nobel Laureate Maathai Links Environment to Peace, Democracy."
24. *Shamba* is a Swahili word that means *field.* The term *shamba system* is generally defined in Kenya as a form of agro forestry through which farmers are encouraged to cultivate crops on previously clear-cut forest land on condition that

they replant the forest trees. After three years of cultivating, the trees are expected to be grown enough to overshadow the agricultural crops. The farmer then moves out of the allocated plot and is eligible for another plot to be cleared.

25. Muthuki, "Challenging Patriarchal Structures," 83–91.

26. MacDonald, "The Prize Is Right."

27. MacDonald, "The Prize Is Right."

28. Okome Olúfúnké Mojúbáolú, "The Relevance of the AMISTAD to Ralph Bunche, Wangari Maathai and Barack Obama."

29. The Green Belt Movement, "The Nobel Peace Prize."

30. Muthuki, "Challenging Patriarchal Structures, 83–91."

31. Maathai, Nobel lecture.

32. United Nations Environment Programme, "The Environment in the News."

33. GO Committee on Education, "Beijing Platform for Action: Women and the Environment."

34. Jiggins, *Changing the Boundaries;* Economic and social development department, "Gender and Development Plan of Action."

35. NGO Committee on Education, "Beijing Platform for Action: Women and the Environment."

36. Rosa, "State of the Future 2005: Update on Women's Issues."

37. Norgaard, "Gender Equality and State Environmentalism."

38. Norgaard, "Gender Equality and State Environmentalism."

39. Mills, "Eco-feminism: Women around the World."

40. Cuomo, "Towards Thoughtful Ecofeminist Activism"; Gaard, *Ecofeminism;* Salleh, "Living with Nature"; and Warren, "Feminism and Ecology."

41. Shiva, *Staying Alive.*

42. Etkins, *A New World Order;* Mies, *Patriarchy and Accumulation on a World Scale;* and Shiva, *Staying Alive.*

43. Shiva, *Staying Alive.*

44. Thomas-Slayter and Rocheleau, *Gender, Environment and Development in Kenya.*

45. Dankelman, "Women and Environment: Weaving the Web of Life."

46. MacGregor, *Beyond Mothering Earth.*

47. United Nations Environment Programme, "The Environment in the News."

48. Barbash and Wax, "Kenyan Woman Wins Nobel Peace Prize."

49. Barbash and Wax, "Kenyan Woman Wins Nobel Peace Prize."

50. Hunma, "Alternative Avenues in the Name of Peace: A Rhetorical Analysis of Nobel Lectures of Wangari Maathai, Rigoberta Menchu and Aung San Suu Kyi."

51. The Ogiek People /ecoterra, "Nobel Laureate Maathai Links Environment to Peace, Democracy."

52. Tyler, "Kenyan Awarded Nobel Peace Prize."

53. MacGregor, *Beyond Mothering Earth.*

54. United Nations Environment Programme, "The Environment in the News."

55. Fadahunsi, "Kenyan Ecologist Awarded 2004 Nobel Peace Prize," and Janet Muthuki "Challenging Patriarchal Structures."

56. Maathai, *The Green Belt Movement.*

57. Dankelman, "Women and Environment: Weaving the Web of Life."

58. Women and Population Division Sustainable Development Department FAO, "Participation and Information: The Key to Gender-responsive Agricultural Policy."

59. NGO Committee on Education, "Beijing Platform for Action: Women and the Environment."

60. Pearson, *Gender Matters in Development.*

61. Dankelman, "Women: Agents of Change for a Healthy Environment."

62. Chiuri et al., "Effects of Change in Land Tenure and Resource Management on Gender Relations," and Morna, "Democracy and the Environment."

63. Norgaard, "Gender Equality and State Environmentalism."

64. Wangari et al., "Gendered Visions for Survival."

65. Norgaard, "Gender Equality and State Environmentalism."

66. Boserup, *Women's Role in Economic Development;* and Maathai, *The Green Belt Movement.*

67. Mikell, *African Feminism;* Mies, *Patriarchy and Accumulation on a World Scale;* and Shiva, *Staying Alive.*

68. Oyewumi, *The Invention of Women,* and Muthuki, "Challenging Patriarchal Structures."

69. Muthuki, "Challenging Patriarchal Structures," 83–91.

70. Thomas-Slayter and Rocheleau, *Gender, Environment and Development in Kenya.*

71. Emmons, "The Development of Degradation and Impoverishment."

72. Emmons, "The Development of Degradation and Impoverishment."

73. Emmons, "The Development of Degradation and Impoverishment."

74. Wamalwa, *Indigenous Knowledge and Ecological Management.*

75. Emmons, "The Development of Degradation and Impoverishment."

76. Chandler and Njoki, "Indigenous Gendered Spaces."

77. Emmons, "The Development of Degradation and Impoverishment."

78. Chandler and Njoki, "Indigenous Gendered Spaces."

79. Chandler and Njoki, "Indigenous Gendered Spaces."

80. Chandler and Njoki, "Indigenous Gendered Spaces"; and Emmons, "The Development of Degradation and Impoverishment."

81. Emmons. "The Development of Degradation and Impoverishment."

82. Mikell, *African Feminism.*

83. Emmons. "The Development of Degradation and Impoverishment."

84. Emmons. "The Development of Degradation and Impoverishment."

85. Maathai, *The Green Belt Movement.*

86. Mies, *Patriarchy and Accumulation on a World Scale.*

87. Shiva, *Staying Alive.*

88. Merchant, *The Death of Nature.*

89. Emmons, "The Development of Degradation and Impoverishment"; and Oyewumi, *The Invention of Women.*

90. Kanogo, *Women and Environment in History.*

91. Emmons, "The Development of Degradation and Impoverishment."

92. Phanice Sella Webi, "Wood Fuel Use and Its Socio-economic and Ecological Effects."

93. Bates, *Beyond the Miracle of the Market;* Chiuri, "Effects of Change in Land Tenure and Resource Management on Gender Relations and the Subsequent Changes on Highland Ecosystems"; and Chandler and Njoki, *Indigenous Gendered Spaces.*

94. Chandler and Njoki, *Indigenous Gendered Spaces.*

95. Chandler and Njoki, *Indigenous Gendered Spaces.*

96. Riley and Brokensha, *The Mbeere in Kenya;* and Phanice Sella Webi, "Wood Fuel Use and Its Socio-economic and Ecological Effects."

97. Emmons, "The Development of Degradation and Impoverishment."

98. Chandler and Njoki, *Indigenous Gendered Spaces.*

99. Chandler and Njoki, *Indigenous Gendered Spaces.*

100. Chandler and Njoki, *Indigenous Gendered Spaces.*

101. Emmons, "The Development of Degradation and Impoverishment."

102. Maathai, *The Green Belt Movement.*

103. Emmons, "The Development of Degradation and Impoverishment."

104. Mies, *Patriarchy and Accumulation on a World Scale.*

105. Emmons, "The Development of Degradation and Impoverishment."

106. Kanogo, "Women and Environment in History," and Emmons, "The Development of Degradation and Impoverishment."

107. Mills, "Eco-feminism: Women around the World."

108. Mies, *Patriarchy and Accumulation on a World Scale.*

109. Emmons, "The Development of Degradation and Impoverishment."

110. Ahluwalia, *Post-colonialism and the Politics of Kenya;* Bates, *Beyond the Miracle of the Market;* Berman and Lonsdale, *Unhappy Valley;* and Emmon, "The Development of Degradation and Impoverishment."

111. Etkins, *A New World Order.*

112. Githiji, *Ten Millionaires and Ten Million Beggars;* and Ogot and Ochieng, *Decolonisation and Independence in Kenya.*

113. Berman and Lonsdale, *Unhappy Valley;* Gertzel, *The Politics of Independent Kenya, 1963–8;* Hartmann, "Degradation Narratives"; and Shiva, *Staying Alive.*

114. Chandler and Njoki, "Indigenous Gendered Spaces."

115. In agriculture "monoculture" describes the planting and cultivation of a single species over a substantial area, or the practice of relying on a very small number of species for commercial agriculture. The rise of monoculture in modern agriculture has been the result of improved technology, with machinery for tilling, planting, pest control and harvesting generally being much cheaper and more efficient than human labor. Monocultures are criticized by ecologists because of their susceptibility to disease and insects, the large amount of chemical inputs often required to sustain them, and lower biodiversity (diversity of life in all its forms and at all levels of organization). Ecological movements advocate for more diverse cropping systems.

116. Maathai, *The Green Belt Movement,* and Warren, "Feminism and Ecology."

117. Chandler and Njoki, "Indigenous Gendered Spaces."

118. Allen and Barnes, "The Causes of Deforestation in Developing Countries," 163–84.

119. Allen and Barnes, "The Causes of Deforestation in Developing Countries," 163–84.

120. Chandler and Njoki, "Indigenous Gendered Spaces."

121. Fortmann, *Gendered Knowledge.*

122. Chandler and Njoki, "Indigenous Gendered Spaces."

123. Riley and Brokensha, *The Mbeere in Kenya,* and Chandler and Njoki, "Indigenous Gendered Spaces."

124. Dankelman and Davidson, *Women and Environment in the Third World.*

125. Chandler and Njoki, "Indigenous Gendered Spaces."

126. Maathai, *The Green Belt Movement.*

127. Maathai, *The Green Belt Movement.*

128. Dankelman and Davidson, *Women and Environment in the Third World;* and Maathai, *The Green Belt Movement.*

129. Thomas-Slayter and Rocheleau, *Gender, Environment and Development in Kenya.*

130. Oyewumi, *Invention of Women.*

131. Dankelman, "Women and Environment: Weaving the Web of Life."

132. Dankelman and Brokensha, *Women and Environment in the Third World.*

133. Chiuri et al., "Effects of Change in Land Tenure and Resource Management on Gender Relations and the Subsequent Changes on Highland Ecosystems."

134. Dankelman and Brokensha, *Women and Environment in the Third World.*

135. Dankelman and Brokensha, *Women and Environment in the Third World.*

136. Chiuri et al., "Effects of Change in Land Tenure and Resource Management on Gender Relations and the Subsequent Changes on Highland Ecosystems."

137. Muthuki, "Challenging Patriarchal Structures," 83–91.

138. Maathai, *The Green Belt Movement.*

139. Maathai, *The Green Belt Movement.*

140. Maathai, *The Green Belt Movement.*

141. Maathai, *The Green Belt Movement.*

142. Maathai, *The Green Belt Movement.*

143. Maathai, *The Green Belt Movement.*

144. Maathai, *The Green Belt Movement.*

145. Muthuki, "Challenging Patriarchal Structures," 83–91.

146. Maathai, *The Green Belt Movement.*

147. Cravey, "Engendering the Latin American State," 523–42, and Norgaard, "Gender Equality and State Environmentalism."

148. Muthuki, "Challenging Patriarchal Structures," 83–91.

149. Muthuki, "Challenging Patriarchal Structures," 83–91.

150. Fadahunsi, "Kenyan Ecologist Awarded 2004 Nobel Peace Prize," and The Hunger Project, "Professor Wangari Muta Maathai 1991 Africa Prize Laureate."

151. Maathai, *The Green Belt Movement.*

152. Maathai, Nobel lecture.

153. Maathai, *The Green Belt Movement,* and The Ogiek people/ ecoterra, "Nobel Laureate Maathai Links Environment to Peace, Democracy."

154. Maathai, *The Green Belt Movement.*

155. Mikell, *African Feminism.*

156. Mikell, *African Feminism.*

157. Muthuki, "Challenging Patriarchal Structures," 83–91.

158. Maathai, *The Green Belt Movement.*

159. Bond, *Unsustainable South Africa.*

160. Braidotti et al., *Women, the Environment and Sustainable Development.*

161. Chandler and Njoki, "Indigenous Gendered Spaces."

162. United Nations Women, "The Centrality of Gender Equality and Empowerment of Women for Sustainable Development."

163. Mohammed, *Alternative Development Strategies for Africa,* and Steady, *Women and Children First.*

164. Jiggins, *Changing the Boundaries.*

165. Loots and Witt, "Beijing+10," 52–62.

166. Pambazuka News, "Nobel Peace Prize Winner Addresses the African Diaspora," and Muthuki, "Challenging Patriarchal Structures."

167. Maathai, *The Green Belt Movement.*

168. Maathai, *The Green Belt Movement,* and Mies, *Patriarchy and Accumulation on a World Scale.*

169. Maathai, *The Green Belt Movement;* and Mies, *Patriarchy and Accumulation on a World Scale.*

170. Beneria, *Gender, Development and Globalization.*

171. Agonafer, *Contending Theories of Development in the Contemporary International Order/Disorder;* Ahluwalia, *Post-colonialism and the Politics of Kenya;* Bates, *Beyond the Miracle of the Market;* and Ogot and Ochieng, *Decolonisation and Independence in Kenya, 1940–1993.*

172. Muteshi, "African Women's Engagement with the United Nations."

173. Maathai, *The Green Belt Movement.*

174. Maathai, *The Green Belt Movement,* and Wangari Maathai, Nobel lecture.

175. Maathai, Nobel lecture.

176. Nepal, "How Movements Move," 24–29.

177. Nepal, "How Movements Move," 24–29.

Bibliography

Agonafer, Mulugeta. *Contending Theories of Development in the Contemporary International Order/Disorder: Lessons from Kenya and Tanzania.* Lanham, MD: University Press of America, 1994.

Ahluwalia, Pal D. *Post-colonialism and the Politics of Kenya.* New York: Novas Science Publishers, 1996.

Allen, Julia C., and Douglas F. Barnes. "The Causes of Deforestation in Developing Countries." *Annals of the Association of American Geographers* 75, no. 2 (1985): 163–84.

Barbash, Fred and Emily Wax. "Kenyan Woman Wins Nobel Peace Prize." *The Washington Post,* 2004. http://www.rawfoodinfo.com/articles/art_wangarimaathaiwins.html (Accessed January 7, 2013).

Bates, Robert H. *Beyond the Miracle of the Market: The Political Economy of Agrarian Development in Kenya.* Cambridge: Cambridge University Press, 1989.

Beneria, Lourdes. *Gender, Development and Globalization: Economics As If All People Mattered.* New York: Routledge, 2003.

Berman, Bruce J., and Lonsdale John. *Unhappy Valley: Conflict in Kenya and Africa.* London: James Currey, 1992.

Bond, Patrick. *Unsustainable South Africa: Environment, Development and Social Protest.* Pietermaritzburg: University of Natal Press, 2002.

Boserup, Esther. *Women's Role in Economic Development.* New York: St. Martin's Press, 1970.

Braidotti, Rosi, Ewa Charkiewicz, Sabine Häusler, and Saskia Wieringa. *Women, the Environment and Sustainable Development: Towards a Theoretical Synthesis.* London: Zed Books, 1994.

Bramwell, Anna. *Ecology in the 20th Century: A History.* New Haven, CT: Yale University Press, 1989.

Chandler, Debra, and Wane Njoki. "Indigenous Gendered Spaces: An Examination of Kenya." *Jenda: A Journal of Culture and African Women Studies* 2, no. 1 (2002). http://www.jendajournal.com/vol2.1/chandler-wane.html (Accessed August 20, 2011).

Chiuri, Wanjiku. "The Effects of Change in Land Tenure and Resource Management on Gender Relations and the Subsequent Changes on Highland Ecosystems: A Case Study of Iruri-Kiamariga Community in the Nyeri District, Kenya." PhD diss., University of Waterloo, 1996.

Cline-Cole, Reginald A., C. Main, and Janet E. Nichol. "On Fuelwood Consumption, Population Dynamics and Deforestation in Africa." *World Development* 18, no. 4 (1990): 23–37.

Cuomo, Christine. "Towards Thoughtful Ecofeminist Activism." In *Ecological Feminist Philosophies,* edited by Karen Warren. Indianapolis: Indiana University Press, 1996.

Cravey, Altha. "Engendering the Latin American State." *Progress in Human Geography* 22, no. 4 (1998): 523–42.

Dankelman, Irene. "Women: Agents of Change for a Healthy Environment." *Gender Digest* 22 (2009). http://www.ww05.org/english3/speech/5.2.IreneDankelman.pdf (Accessed January 9, 2013).

Dankelman, Irene. "Women and Environment: Weaving the Web of Life." The opening speech at WECF's Beijing+10 side-event at the United Nations headquarters in New York on March 5, 2005. http://www.wecf.eu/english/articles/2005/03/Beijing10sideevent.php (Accessed January 7, 2013).

Dankelman, Irene, and Joan Davidson. *Women and Environment in the Third World: Alliance for the Future.* London: Earthscan Publications Limited, 1988.

Dobson, Andrew. *Green Political Thought: An Introduction.* New York: Routledge, 1992.

Economic and Social Development Department, FAO Corporate Document Repository. "Gender and Development Plan of Action," 2003. http://www.fao.org/docrep/005/Y3969E/y3969e04.html. (Accessed January 07, 2013).

Ekins, Paul. *A New World Order: Grassroots Movements for Global Change.* London: Routledge, 1992.

Emmons, Gavin. "The Development of Degradation and Impoverishment: Neocolonialism and the Crisis of People and the Environment in East Africa." MA diss., Lewis and Clark College, 1996. http://www.lclark.edu/~soan/abstract.html (Accessed throughout August 2011).

Fadahunsi, O. "Kenyan Ecologist Awarded 2004 Nobel Peace Prize." *The African* (2005).http://www.africanmag.com/content/african_back_article_.aspx?article=223 (Accessed August 15, 2011).

Fortmann, Louise. "Gendered Knowledge: Rights and Space in Two Zimbabwe Villages, Reflections on Methods and Findings." In *Feminist Political Ecology, Global Issues and Local Experiences,* edited by Dianne Rocheleau, Barbara Thomas-Slayter and Esther Wangari. New York: Routledge, 1996.

Foundation for Sustainable Development. "Environmental Sustainability Issues in Kenya." (2010). http://www.fsdinternational.org/devsubject/environment/kenyaissues (Accessed January 6, 2013).

Gaard, Greta. *Ecofeminism: Women, Animals, Nature.* Philadelphia: Temple University Press, 1993.

Gaard, Greta. *Ecological Politics: Ecofeminists and the Greens.* Philadelphia: Temple University Press, 1998.

Gertzel, Cherry J. *The Politics of Independent Kenya, 1963–8.* Evanston, IL: Northwestern University Press, 1970.

Gilson, D. "Root Causes: An Interview with Wangari Maathai." *Mother Jones* (2005). http://www.motherjones.com/news/qa/2005/01/wangari_maathai.html (Accessed August 15, 2011).

Githiji, Mwangi. 10 *Millionaires and 10 Million Beggars: A Study of Income Distribution and Development in Kenya.* Aldershot: Ashgate, 2000.

The Green Belt Movement. "The Nobel Peace Prize." (n.d.) http://greenbeltmovement.org/wangari-maathai/the-nobel-peace-prize (Accessed January 5, 2013).

Hartmann, Betsy. "Degradation Narratives." IHDP Update, *Newsletter of the International Human Dimensions Programme on Global Environmental Change,* Number 4/2002. http://www.ihdp.unibonn.de/html/publications/update/update02_04/IHDPUpdate02_04_Hartmann.htm (Accessed August 8, 2011).

The Hunger Project. "Professor Wangari Muta Maathai 1991 Africa Prize Laureate: The Founder of the Green Belt Movement." (n.d.) http://www.thp.org/what_we_do/key_initiatives/honoring_africa_leadership/laureate_list/professor_maathai (Accessed January 8, 2013).

Hunma Bye, Aditi, "Alternative Avenues in the Name of Peace: A Rhetorical Analysis of Nobel Lectures of Wangari Maathai, Rigoberta Menchu and Aung San

Suu Kyi." (2007). http://ebookbrowse.com/aditi-hunma-women-nobel-peace-prize-addresses-pdf-d272972575 (Accessed January 8, 2013).

Jiggins, Janice. *Changing the Boundaries: Women Centred Perspectives on the Population and Environment.* Washington, DC: Island Press, 1994.

Kanogo, Tabitha. "Women and Environment in History." In *Groundwork: African Women as Environmental Managers,* edited by Shanyisa A. Khasiani. Nairobi: Acts Press, 1992.

Loots, Lliane, and Harold Witt. "Beijing +10: Women and the Environment—How Close Are We to Democracy?" *Agenda* 64 (2005): 52–62.

Maathai, Wangari. *The Green Belt Movement: Sharing the Approach and the Experience.* New York: Lantern Books, 2004.

Maathai, Wangari. "Nobel Lecture." December 10, 2004. http://nobelprize.org/peace/laureates/2004/maathai-lecture-text.html (Accessed October 20, 2011).

MacDonald, Mia. "The Prize is Right." *Satya* (2005). http://www.satyamag.com/jan05/macdonald.html (Accessed January 4, 2013).

MacGregor, Sherilyn. *Beyond Mothering Earth: Ecological Citizenship and the Politics of Care.* Vancouver, UBC Press, 2006.

Merchant, Carolyn. *The Death of Nature: Women, Ecology and the Scientific Revolution.* San Francisco: Harper and Row, 1980.

Mies, Maria. *Patriarchy and Accumulation on a World Scale: Women in the International Division of Labour.* London: Zed Books, 1998.

Mikell, Gwendolyn. *African Feminism: The Politics of Survival in Sub-Saharan Africa.* Philadelphia: University of Pennsylvania Press, 1997.

Mills, Coeta. "Eco-feminism: Women around the World." 1989. http://www.holysmoke.org/fem/fem0097.htm (Accessed January 7, 2013).

Mohammed, Suliman. *Alternative Development Strategies for Africa: Environment and Women.* London: The Institute for Africa Alternatives, 1991.

Morna, Colleen. "Democracy and the Environment." *The Power to Change: Women in the Third World Redefine their Environment, Women's Feature Service.* London: Zed Books, 1992.

Muteshi, Jacinta. "African Women's Engagement with the United Nations." *AWID* (2006). http://www.awid.org/eng/Library/African-Women-s-Engagement-with-the-Unhp?stid=1605.Site (Accessed June 14, 2012).

Muthuki, Janet. "Challenging Patriarchal Structures. Wangari Maathai and the Green Belt Movement in Kenya." *Agenda* 69 (2006): 83–91.

Muthuki, Janet. "Rethinking Ecofeminism: The Case of Wangari Maathai and the Green Belt Movement in Kenya." MA diss., University of KwaZulu-Natal, 2005.

Nepal, Padam. "How Movements Move: Evaluating the Role of Ideology and Leadership in Environmental Movement Dynamics in India with Special Reference to the Narmada Bachao Andolan." *Hydro Nepal* 4 (2009): 24–29.

NGO Committee on Education. "Beijing Platform for Action. Chapter IV.K. Women and the Environment." http://www.un-documents.net/bpa-4-k.htm (Accessed January 6, 2013).

Norgaard, Kari. "Gender Equality and State Environmentalism." *Gender & Society* (2005): 506–22. http://www.whitman.edu/environmental_studies/faculty/GS19.pdf (Accessed January 8, 2013).

The Norwegian Nobel Committee. "The Nobel Peace Prize 2004. Press release Oslo." October 8, 2004. http://www.nobelprize.org/nobel_prizes/peace/laureates/2004/press.html (Accessed January 10, 2013).

The Ogiek people/Ecoterra Intl. "Nobel Laureate Maathai Links Environment to Peace, Democracy." (2005). http://www.ogiek.com/indepth/maathai-website.htm (Accessed January 5, 2013).

Ogot, Bethwell A., and William R. Ochieng. *Decolonisation and Independence in Kenya, 1940–1993.* London: James Currey, 1995.

Okome Olúfúnké Mojúbáolú. "The Relevance of the AMISTAD to Ralph Bunche, Wangari Maathai and Barack Obama." Public Lecture Prepared for Presentation for the Seventh Amistad Lecture, Central Connecticut State University New Britain, Connecticut, February 23, 2010. *Africa Update Newsletter* 17, no. 1 (2010). http://webccsu.edu/astudy/upd17–1.html (Accessed January 10, 2013).

Oyewumi, Oyeronke. *The Invention of Women: Making an African Sense of Western Gender Discourse.* Minneapolis: University of Minnesota Press, 1997.

Pambazuka News 214- Addendum. "Nobel Peace Prize Winner Addresses the African Diaspora." (2005). http://www.pambazuka.org/index.php?id=28870. (Accessed August 14, 2011).

Pearson, Ruth. "Gender Matters in Development." In *Poverty and Development in the 1990s,* edited by Allen Tim and Thomas Alan. Oxford: Oxford University Press, 1993.

Phanice Sella Webi. "Wood fuel Use and its Socio-economic and Ecological Effects." In *Human Landscape Ecology* (MNFEL 330/RFEL 3031). Selected Term Papers 2003/2004, edited by Kerstin Polthoff (2005). http://www.svt.ntnu.no/geo/Doklager/Acta/Human%20landscape%20ecology.pdf (Accessed January 8, 2013).

Ponting, Clive. *A Green History of the World.* London: Sinclair-Stevenson, 1991.

Porritt, Jonathon. *Seeing Green: The Politics of Ecology Explained.* London: Blackwell, 1991.

Riley, Bernard W., and David Brokensha. *The Mbeere in Kenya, Volume one.* New York: University Press of America, 1988.

Rodda, Annabel. *Women and the Environment: Women and World Development Series.* London: Zed Books, 1991

Rosa, Alegria. "State of the Future: Update on Women's Issues." Report to the Millennium Project Planning Committee Meeting July 2005. http://www.millennium-project.org/millennium/mppc-0705/Brazil-Women%20Study.doc (Accessed December 27, 2012).

The Royal Forums. Address by Professor Ole Danbolt Mjøs, Chairman of the Norwegian Nobel Committee, Oslo, December 10, 2004. http://www.theroyalforums.com/forums/f16/nobel-peace-prizie-award-2004-a-4305.html (Accessed January 5, 2013).

Salleh, Ariel. "Living with Nature: Reciprocity or Control." In *Ethics of Environment and Development,* edited by Ronald and Joan Engel. Tucson: University of Arizona Press, 1988.

Shiva, Vandana. *Staying Alive: Women, Ecology and Development.* London: Zed Books, 1988.

Steady, Filomina. *Women and Children First: Environment, Poverty and Sustainable Development.* Rochester: Schenkman Books Inc, 1993.

Thomas-Slayter, Barbara, and Rocheleau Dianne. *Gender, Environment and Development in Kenya: Grassroots Perspectives.* London: Lynne Rienner Publishers, 1995.

Timberlake, Lloyd. *Africa in Crisis: The Causes and Cures of Environmental Bankruptcy.* London: Earthscan, 1985.

Tyler, Patrick. "Kenyan Awarded Nobel Peace Prize." *The New York Times,* October 9, 2004. http://www.nytimes.com/2004/10/08/world/08iht-nobel.html?_r=1 (Accessed July 20, 2011).

United Nations Environment Programme. "The Environment in the News." December 14, 2004. http://www.unep.org/cpi/briefs/Brief14Dec04.doc (Accessed January 6, 2013).

United Nations Women. "The Centrality of Gender Equality and Empowerment of Women for Sustainable Development." http://www.unwomen.org/wp-content/uploads/2011/11/Rio+20-UN-Women-Contribution-to-the-Outcome-Document.pdf (Accessed January 8, 2013).

Wangari, Esther, Barbara Thomas-Slayter, and Rocheleau Dianne. "Gendered Visions for Survival: Semi-arid Regions in Kenya." In *Feminist Political Ecology: Global Issues and Local Experiences,* edited by Esther Wangari, Barbara Thomas-Slayter and Rocheleau Dianne. New York: Routledge, 1996.

Wamalwa, Betty. "Indigenous Knowledge and Ecological Management." In *Gaining Ground: Institutional Innovations in Land-use Management in Kenya,* edited by Amos Kiriro and Juma Calestous. Nairobi: ACTS Press, 1991.

Warren, Karen J. "Feminism and Ecology: Making Connections." *Environmental Ethics* 9, no. 1 (1987): 3–20.

Women and Population Division, Sustainable Development Department FAO. "Participation and Information: The Key to Gender-responsive Agricultural Policy." Document for the High-Level Consultation on Rural Women and Information Rome, October 4–6, 1999. http://www.fao.org/docrep/X2950E/X2950E00.htm (Accessed January 8, 2013).

Chapter 9

Green Movements in Nigeria

John Agbonifo

Introduction

Environmentalism—environmental meanings, beliefs, values, and behaviors—is far from a recent phenomenon in Africa. The anticolonial coalition included environmental movements.[1] In late-19th-century West Africa, disputes between chiefs and the colonial state over forests management prompted the former to team up with proto-nationalists such as Casely-Heyford.[2] Following national independence, such movements, which were later seen as rival currents by factions of the ruling elites, were demobilized in the name of modernization and national unity. Social movements, including environmental movements, especially those that challenged the order of things, were crushed. Beyond such radical action, the process of modernization itself crippled environmentalism in Nigeria. As the dominant and organizing discourse of progress or social transformation, modernization had the effect of subjugating existing knowledge about the environment, environmental concerns, and consciousness. The subjugation found rationalization and justification in the grammar of resource development. Modernization was construed as the vehicle of the much-awaited transition from poverty, disease, and backwardness to progress, economic plenty, and happiness. Any challenge or opposition in the way of modernization was vilified as ignorance, backwardness, and support for national disunity.

Oil development in Nigeria provides the paradigmatic context for the examination of environmentalism in the country. The oil industry emerged within a colonial context in which sole exploratory rights covering the entire land surface of the country was given to Shell. Both the colonial state and Shell were commercial entities. They were more interested in global capitalist accumulation for the benefit of their home country than Nigeria. Within the exploitative mindset of colonialism, the negative impacts of commercial enterprise mattered little. Seeking to safeguard its business interests, the colonial state introduced laws that had little benefit for the country. Some of the laws were negotiated during the process of transferring power to independence leaders. Thus, it took almost a decade after independence before the colonial law governing the oil industry was altered. And even then the change effected was not significantly different from the status quo ante.

There is a paucity of independent and authoritative studies of the impact of the oil industry on which both local communities and Shell/state agree. The World Bank commissioned a major study of the Niger Delta, and Shell initiated and funded another study, the Niger Delta Environmental Survey. Both studies, however, focus broadly on the entire region and they do not account for the social and cultural impacts of the industry. In this context, claims and counterclaims over the impact of oil activities characterize community-Shell relations. The absence of a commonly accepted benchmark against which to judge claims points to a deeper problem. The oil industry in Nigeria did not, and still has not, produced an evaluative appraisal of the value of oil development, its overall effects, tangible and intangible, on host communities, their environment, and to what extent the benefits of oil exploitation outweighs its effects, if at all.

Explanations for the gap can be located in the uncritical acceptance in the 1950s and 1960s of the ruling development ideology in which resource export, particularly mineral exports, was seen as the surest way to industrial take-off and prosperity. The mere discovery of crude oil was greeted with excitement. Oil was seen by both the nationalist leaders and local peoples as the panacea to poverty and underdevelopment. That development, however, has a dark side was hardly contemplated. Indeed, so remote was any consciousness of the negative impact of industry that debates about the role of Foreign Direct Investment (FDI) and Transnational Corporation (TNCs) focused exclusively on whether or not they can foster industrialization in the periphery. Inattention to the environmental impact of industry suggests a narrow-minded perception of development as an entirely positive process.

Nevertheless, the petroleum industry has transformed the nature, dynamics, and trajectory of Nigeria's economic development. This is reflected in the fact that while the first national development plan was based on external sources of funds, the second and third plans, which were more ambitious, were based on internally generated revenue from oil.[3] The dark side of oil has, however, bared its fangs. Environmental pollution and uneven distribution of the externalities of oil production have given rise to contentious politics. Faced with the possibility of foreign exchange earnings for socioeconomic transformation, in a context of scarce capital resources, the state has given little attention to environmental well-being. That is the case because the state tends to sacrifice environmental quality for the benefits of oil development—increasing national per capita income and improved standard of living.[4] This has had the effect of provoking grievances and grassroots contentious mobilization against the industry.

In this chapter, an attempt is made to sketch what environmentalism means in Africa and the ways in which such meanings have been misunderstood, through careful delineation of an African environmental ethic. The ethics have been obscured by secular and economic reductionisms, and a practical example is employed to expose such reductionisms. It makes a case for varieties of environmentalisms in order to guide against the common tendency to substitute a part for the whole. The remaining part of the chapter is organized into four sections. The first section outlines an African environmental ethic. The second section fleshes out the historical background to environmental exploitation and environmentalism in Nigeria. Following that is an examination of the evolution of modern environmentalism and the varieties or forms of environmentalisms in the country. Thereafter, we take the Ogoni-State/Shell conflict as a case study to highlight the chapter's main arguments.

African Environmental Ethics and Reductionism

Western models and approaches to environmental movements have proven inadequate in explaining third world environmentalism.[5] Some African scholars and environmentalists have contributed to the confusion by their partial understanding and portrayal of the environment, and, inevitably, their inadequate grasp of African environmentalism. For example, Ogungbemi argues that African environmentalism rests on the ethics of "nature-relatedness." The author, however, reduces nature to a material actor in its relationship with human beings. He fails to reckon that nature has a spiritual dimension as it harbors deities, ghosts, and gods. The author provides justification for the critique when he claims that natural resources do not have any religious affinity. The ethics of "nature-relatedness" merely emphasizes the need for peaceful coexistence between nature and human beings, and the importance of treating nature as a worthy actor in itself.[6] Such understanding rehashes Western conservationist and preservationist philosophy.

Godfrey Tangwa counterposes the African environmental worldview to the Western worldview, which is predominantly anthropocentric and individualistic.[7] The author locates precolonial traditional African environmental ethics in African metaphysics, or "eco-biocommunitarianism," the "recognition and acceptance of inter-dependence and peaceful coexistence between earth, plants, animals and humans."[8] The metaphysical outlook explains the benign relations between humans, animals, and inanimate objects and supra-natural forces.[9]

Within African metaphysics, the dichotomy between "plants, animals and inanimate things, between the sacred and the profane, matter and spirit, the communal and the individual, is a slim and flexible one."[10] Against such a worldview, it becomes clearer as to why African people believe that human beings can turn into animals, plants, and unseen forces, and vice versa. Such a view portends important implications for how traditional African people approach and treat the environment. Ojomo argues that the thin boundary between the holy and the profane, the human and the supra-human, and the communal and the individual shaped the traditional African worldview of "live and let live."[11] Thus, eco-biocommunitarianism is not the metaphysics of domination, consumerism or greed, but ideas and claims rooted in myths and taboos that serve to conserve ecological balance. Notwithstanding the foregoing insights, two major theoretical misunderstandings have limited scholarly comprehension of environmentalism in Africa. These are, for want of better terms, economic and secular reductionisms.

Economic Reductionism

Not a few Western scholars begin their analysis of environmentalism in non-Western societies from the premise of poverty or economic gain. The poor according to them are more concerned about bread-and-butter issues than with environmental issues. As a result, Guha and Martinez-Alier argue that the "environmentalism of the poor" differs from environmentalism in the relatively affluent countries of Western Europe and North America.[12] The authors construe the environmentalism of the wealthy, developed world as largely post-materialist. It is more concerned with the rights of the "nonhuman" world than with bread and butter. Guha and Martinez-Alier have, however, been accused of unfair and inaccurate representation of environmentalism in many parts of the world.[13]

More specifically, William Reno employs the economic lens to x-ray the environmental protests of the Ogoni people against Shell in Nigeria. He claims that far from a genuine attempt to overthrow the system of exploitation, the protest is no more than a cunning attempt to secure a place in the patronage network.

> In Nigeria, it is possible that opposition figure and former governor under military rule Ken Saro-Wiwa attempted to use local muscle to force his way back into the governor's office under a proposed transition to civilian rule. This member of the Nigerian political class discovered that he could tap into environmental rhetoric (and real grievances) of people in oil-producing regions of Nigeria. This gave Saro-Wiwa the possibility of mobilizing outside nonstate actors to pressure Nigeria's military government to give more resources to his region (and the local government that Saro-Wiwa perhaps wished to head).[14]

Reno echoes the analytical distinction in social theory about how the grammar of recognition struggles marginalizes and displaces redistribution struggles rather than supplement and enrich them.[15] Swanson took exception to Fraser's separation between economics, politics, and culture because no relationship or practice is entirely purely economic or cultural.[16]

Given that every struggle against injustice is a demand for both redistribution and recognition,[17] the distinction between redistribution and recognition is based on a simplified and mistaken view of social recognition.[18] A problem with Reno's economic reductionism is the failure to appreciate that the so-called economic is entangled with the cultural. From the African metaphysical worldview, environmentalism in Nigeria is as much about ethics, gods, and totems as it is about daily survival. Contrary to Guha and Martinez-Alier's suggestion, poor groups in Nigeria exhibit a strong interest in environmental matters.[19] Indeed, the growth in the number of environmental interest groups in the country since the early 1970s shows the organized effort by citizens to combat environmental problems.

Secular Reductionism

Much of the literature on environmentalism in Nigeria limits environmental concerns to the physical environment. There is little or no regard for the otherworldly components of the environment, including the gods, spirits and ancestors that local peoples invest the environment with. Many societies hold a worldview that includes beliefs in the existence of an invisible nonmaterial world and "a causal relation between human events and supernatural determinants."[20] Hobsbawm introduced a duality of religion and politics when he argued that 20th-century peasants' social movements are archaic because their actions are very often organized in kinship or religious terms, and entangled with supernatural actors.[21]

Chakrabathy criticizes the ontological assumption that "the human is ontologically singular, that gods and spirits are in the end 'social facts,' that the social somehow exists prior to them."[22] He emphasizes that there is no known society in which humans have existed without relating with gods and spirits; gods and spirits are "existentially coeval with the human," and "the question of being human involves the question of being with gods and spirits."[23] Land, forests, rivers, streams have spiritual significance. They are in some places totems, and in others gods. As such human relations with them are controlled through cultural norms and rules. Sacred forests were not sacred because people needed them for economic survival. They were protected because the people understood them as entities in their own sacred right and therefore to be conserved. More than that, they were to be conserved because their degradation implied human or societal degradation given that the environment was merely the external embodiment of society.

Struggles over environmental resources are simultaneously struggles over cultural meanings. The process of resource distribution involves simultaneous struggles over meaning as well as struggles over surplus.[24] The symbolic dimension of struggles over environmental resources has been deemphasized.[25] What makes oil-induced environmental degradation of consequence here is the fact that it happens in place; a portion of space invested with meaning by some people and to which they have become attached. A nonplace approach to the Niger Delta sees nothing but a world of vast oil resources and nature. However, the conception of the Niger Delta as a world of places enables a deeper perception; that of worlds of meanings and experiences, connections, and attachment between people and places. In other words, a place perspective frees us from the rationalizing gaze of modernity, forcing on us a decolonized understanding that the so-called nature, far from being pure, is the outcome of both social and natural processes. A worldview founded on the modernist understanding of environment as space, empty and without meaning, does violence to constructed nature and extinguishes the basic coordinates by which people negotiate life.

Background to Environmental Exploitation and Environmentalism

This section focuses on the two major currents, modernization and resource-related law-making, which facilitated environmental exploitation and simultaneously sowed the seed of modern environmentalism in Nigeria.

Both currents emerged in the colonial era when colonized peoples were being exposed significantly to novel ideas and practices over which they had little control. The purpose is to show that environmental exploitation emerged within a given historical context, and that the processes of exploitation gave rise to modern environmentalism. It emphasizes that modernization and oil regulations came into conflict with preexisting ideas and values about the environment, provoking a diverse range of reactions.

People are involved in four varying orders of interactions: local, regional, national, and international. These channels of social, political, and economic relationship Mehretu refers to as "interaction orbits."[26] The local order is critical to rural dwellers and consists of voluntary socioeconomic exchanges conducted within the locality for the socioeconomic security of the family. The local orbit facilitates social cohesion, participation, and cooperative endeavors. Regional orders are not so critical to rural inhabitants. They are about a market for surplus production and purchase of goods unavailable locally. The inbound track of the regional orbit facilitates government intervention in the form of social services, development projects, and deeper bureaucratic reach.

> Reduction of the importance of the local order of interaction and placing subsistence and basic needs decisions on higher order orbits, weakens the political basis of rural inhabitants. When this happens with external pressure, it is referred to as orbital elongation. When the local orbit is weakened by orbital elongation, it becomes easy for outsiders to manipulate rural people and take advantage of their increased uncertainties.[27]

The national orders are farther removed from the daily routines of rural people. The in-bound track of the national order is composed of numerous demands from the state, including decrees, place- and resource-specific policies, laws on what can be said and demanded and what cannot. The international order of spatial interaction is least important to rural people because they know little about how to engage with it.

Modernization of Exploitation

The ideology and practice of modernization theory constituted an inbound international orbit of interactions for which national elites stood as surrogates. The local people had little control over it as it was an imposed ideology about which they knew very little. The modernization paradigm sought to replicate the Western development pattern through the reproduction of

Rostow's stages of economic growth. Accordingly, development planning in Nigeria, beginning in 1949, was embedded in the modernization paradigm. The four national development plans, 1962–1985, adhered to the logic of Rostowian growth.

> The first and second development plans, which were prepared during the period when the emphasis on Rostow's growth theory was at its peak, were primarily concerned with attaining a minimum level of growth as the means for establishing the conditions for the "take-off" of development. According to Stolper, the main architect of the first development plan, "the whole purpose of the planning exercise (was) to aid Nigeria in mobilising as many resources as possible and allocating them for growth." Operating more clearly within the context of the "stages of growth" theory, the second development plan stated that Nigeria had "all the essential ingredients for a quick 'take off.'" The plan therefore placed emphasis "on growth as a pre-condition for a meaningful distribution of the fruits of development" stating that "the 'national cake' must first be baked before it is shared."[28]

The modernizing logic outlived the colonial era and remains central in the development strategies of independent African states and international agencies.[29]

Thus, the economic profitability of a development project should be the determining factor of investment and the size of investment to be channeled into it. Profitability has no ideological ring to it because it simply measures whether a particular investment does or does not lead to an increase in available resources. If it does, the use of resources is defensible; if not, the expenditure cannot be defended. Sectoral investments have to be based on the capacity of the sector to generate profits.

While the oil industry has been profitable in some areas, Madujibeya argues that when focus is removed from monetary contribution to the real economic impact, the industry has made minimal impact on economic development.[30] Focused on the profitability criteria, handlers of the Nigerian economy were persuaded that the benefits of the oil industry would outweigh its cost, and that the benefits of economic growth would trickle down. Thus, the modernization project in Nigeria failed to consider the environmental impact of oil exploration on host communities.

Development economists who greatly influenced developmentalist policies in Nigeria adhered strictly to the Rostowian "stages of growth" or the linear conception of development. In order to achieve high mass consumption and in the process become like the West, they recommended "policies that promoted authoritarian and accumulationist state."[31] As a result, the development strategies were preoccupied with ways in which the state could mobilize the resources needed for the Rostowian take-off stage. In the process, ancestral groves, forbidden forests, and sacred places were destroyed and taken over by the oil industry in the name of a better future, even as the ideas and beliefs that informed the creation and protection of such places were undermined and crushed.

Legalization of Exploitation

Some of the critical elements of the national inbound track that have shaped the life-chances of local communities in the Niger Delta include the Petroleum Law and the Land Use Decree. Not unlike colonial conservationism, which facilitated appropriation of Africa's natural resources, state laws have been used to expropriate the resources of the Niger Delta.[32] A major contributor to environmental inequality in the region is the system of state laws regulating the energy sector that are in place. Thus, environmental inequality is a function of unequal power relations. The Land Use Act 1978 in particular effectively removes local community control over land, making it easy for oil companies to acquire whatever land they require. Although the Land Use Decree is a national Act, there is no other place in the country that has suffered the degree of land theft and degradation the Niger Delta has endured. The consequence is that the region has had to deal with more levels of decision making about which it knows little. Moreover, its bargaining power dips as it is drawn into higher orbits of interactions by hegemonic fiat.

To gain an insight into how oil-bearing communities are drawn into exploitative relations draped in legal guises, Adeola concisely describes the contents of the national interaction orbits and the mechanism of extraction in the Niger Delta thus:

> (1) an ethnic-centered political dominance, tactically employed to control and exploit the natural resource (wealth) of minority communities for the benefit of the dominant ethnic groups; (2) the alliance of the core ethnic groups, multinational companies, political elites, the military, and the government which generally represses the opportunity structures for the minorities; and (3) massive ecological disruptions and the subsequent destruction of the basic modes of subsistence of the resource-dependent communities of indigenous minorities.[33]

Naanen conceptualizes the mechanism of exploitation as "internal colonialism," involving the strategic pursuit of political dominance by the three major ethnic groups, namely the Hausa-Fulani, the Yoruba, and the Igbo. Political control and numerical preponderance were used by the majority groups to transfer resources from the minority areas to develop the majority areas. The result is the emergence of an economically advantaged and politically powerful core and an impoverished and weak periphery. The multinational oil companies actively aided the domination and peripheralisation.[34] Thus, the transnational ruling class, composed of the ruling class at the global, national and local levels, benefits from the expropriation and degradation of the Ogoni ecology.[35] Such a political economy constitutes environmental injustice and human rights abuse.

The Evolution of Modern Environmentalism in Nigeria

Two sets of events are germane to our understanding of the rise of contemporary environmentalism in Nigeria. The first is the dumping of harmful waste in Koko, Delta state, and the discursive articulation of environmental injustice in the Niger Delta. In 1987, two Italian multinational corporations imported hazardous wastes into Koko. To achieve their aim, the corporations persuaded a certain Sunday Nana into signing an agreement to use his residence for the storage of 18,000 drums of hazardous wastes in return for a monthly payment of $100.[36] Following media exposure, it was discovered that the toxic wastes contained various fatal substances. Those who were exposed to the toxic substances suffered a range of health problems, including Nana himself who died as a consequence.

The publicity given to the development created anxiety among citizens. The dramatic reporting of the devastating impact of toxic waste resulted in many people intuitively perceiving much higher risks to their health and lives than they previously anticipated. Increase in awareness motivates people to seek knowledge in order to reduce their level of ignorance about future potential impacts, enabling environmental groups to gain influence.[37] Thus, following public knowledge of the toxic waste, a range of civil society actions emerged, and organized to promote environmental justice in the region.[38]

The second threshold in the rise of contemporary environmentalism emerged in the early 1990s with the mobilization of oil-bearing communities for decisive political action to end blatant degradation and stake a claim to participation in how the environment is exploited. Communal misgivings about the impact of oil companies on the environment dates back to the early 1960s when Isaac Adaka Boro led a secessionist rebellion against the federal government ostensibly for its complicity in the oil-related environmental degradation of the Niger Delta. In the early 1970s, the Ogoni wrote petitions to the Rivers state government protesting the impact of Shell on their environment. However, it was not until the early 1990s that communities resorted to more direct contentious action to safeguard their environment.

First, the Etche people embarked on a protest against Shell in their community. At the request of Shell, the Nigeria Police invaded the community, killing scores and destroying properties. Few months later, the Ogoni, under the umbrella organization, Movement for the Survival of Ogoni People (MOSOP), began a well-orchestrated campaign against the state and Shell. The Ogoni was visited with state violence, which culminated in the killing of nine Ogoni leaders and the destruction of Ogoni villages. At the point where the MOSOP ebbed, there emerged a cocktail of armed militant groups, claiming ownership of the resources in the Niger Delta and their right to participate in it. The activities of the various groups have drawn worldwide attention to the decades-old assault that oil companies have subjected the Niger Delta environment to.

Varieties of Environmentalisms

The nature and characterization of environmental groups in North America and Europe do not accurately reflect the nature and character of environmentalism in Nigeria. For instance, environmental groups in the United States have been classified into three broad categories, namely oppositional, social justice, and deep ecology groups.[39] Oppositional groups emerge to oppose a single local environmental threat, for example, the "not in my backyard" cases.[40] They do not address a range of environmental issues or act proactively. The environmental justice groups are concerned with environmental problems in poor minority communities. They are single-issue based, concerned primarily with hazards to human health rather than the threat to nature. The third group refers to the deep ecology groups such as Earth First. They retain ecocentric ideologies and employ direct action tactics in their operations.

Like Ogungbemi's ethics of nature-relatedness, the oppositional, social justice and deep ecology threads of American environmentalism do not attribute spiritual nature to the environment. Ecocentrism focuses on the integrity of the ecosystem and the value of species. The ecosphere egalitarianism principle holds that all organisms and entities in the ecosphere are equal in intrinsic value. And all things in the ecosphere have an equal right to live and flourish and attain their own self-realization. The theory of nature's value derives human duties to nature from the value in nature. According to this theory, nature is a subject with creativity, intellect and a capacity to value. These values in nature impose the need for humans to be Earth's moral overseers.

As elsewhere, the environmental movement in Nigeria is not homogenous. It is composed of groups variously interested in wildlife protection, forest conservation, environmental social justice, and environmental education. The movement includes a range of groups at the local, regional, and national levels. Examples are the Nigerian Environmental Society (NES), the Nigerian Conservation Foundation (NCF), the Nigerian Environmental Study and Action Team (NEST), the Green Environmental Movement of Nigeria (GEMINI), Environmental Right Action (ERA), The Nigerian Society for the Protection of the Environment, the Niger Delta Conservation Committee (NDCC) and many other grassroots organizations. Membership in these groups is growing as a result of growing consciousness and concern occasioned by increased visibility of environmental problems and increased mobilization by environmental groups.[41] The environmental groups do not all have the same goals and objectives. Broadly, they are oriented toward promoting environmental awareness, management, and advocacy.[42]

However, environmental groups in Nigeria can be categorized into four groups. First, some environmental groups can be categorized as conservation-oriented environmental groups. Usually elitists, they consist of experts whose job is to carry out investigation into environmental problems, proffer solution and advocate for official policy to stem environmental problems. These groups do not engage in direct illegal public action such as protest. Moreover, they do not challenge the essential validity of the state or society. To the contrary, they work within the system, seeking to effect change through incremental policy change. Their focus is on education based on the assumption that there is a direct link between awareness or education and behavioral change. Very largely, these are single-issue environmental groups, which are either interested in wildlife preservation or forest conservation.

A good example of a conservation group is the Nigerian Conservation Foundation (NCF). Established in 1980, the NCF plays a major role in the promotion of nature conservation and environmental protection. The NCF's core interests include policy advocacy, conservation education, biodiversity surveys, habitat protection, research, and awareness creation through the mass media.[43] Information provided on the group's website has listed the NCF's achievements to include: shaping the establishment of the Federal Environmental Protection Agency (FEPA) and promulgation of the Endangered Species Act No.11 of 1985, ensuring sole management of Okomu Games Sanctuary before its declaration as a national park by the Federal Government. Moreover, it produced the National Conservation Education Strategy for Nigeria in collaboration with the federal ministry of education.

Second, another category of environmental groups includes those that could be described as environmental social justice groups. Such groups are not mainly concerned with wildlife or forest protection like the conservation movements. Their interests lie in protecting minority and vulnerable groups against environmental pollution. Notable in this regard is the Environmental Rights Action (ERA), which came into being in 1993. At inception ERA organized an environmental survey based on which it identified a number of social and environmental problems confronting the country.

ERA utilizes resistance, strategic information access and provision, community empowerment, lobbying, mass education, and litigation as tools to enhance the region's ecological health and well-being.[44] To the group, "resistance" means "the right to say no and mean it in defense of our collective environmental and human rights in order to achieve positive change."[45] Resistance is anchored in direct action: peaceful protests; naming and shaming; blockades; and occupation of the properties or facilities of those responsible for environmental damage. While concerned about environmental problems in the domain of the minority ethnic groups of the Niger Delta, ERA is not oriented toward or driven by a single issue.

Thirdly, there are environmental groups that may be termed peace-building environmental groups. This branch of environmentalism does not blame anyone for the occurrence of an environmental crisis. Adopting a restorative justice approach, their focus is on two questions: who has been harmed, and who is in a position to remedy the harm. In other words, peace-building environmental groups are least interested in blaming or punishing anyone for environmental harm. Through the restorative justice philosophy they seek to promote harmony between polluters and victims of pollution. They work to empower communities to avoid inflicting further damage on themselves and on their environment, and on how best to seek redress for environmental problems. Restorative justice environmentalism is represented by NGOs such as Center for Social and Corporate Responsibility, and Stakeholder Democracy Movement.

A fourth dimension of environmentalism is embodied by local community groups protesting environmental degradation. In the past, these communities could be described as not-in-my-backyard groups because they were focused on single issues such as protest against land grabbing, or environmental pollution. Following years of expropriation and mobilization, these communities have developed the capacity to make the connection between their environmental situation and national and global forces. Thus, while protesting a specific pollution event, they emphasize the need to address the structural causes of socioeconomic marginalization and environmental degradation. For these communities, "environment" means more than physical space; it is the material extension of society, the abode of ghosts and gods. Examples include the Ogoni people, the Ijaws, and many other ethnic groups in the region. A detailed overview of the case of the Ogoni is attempted below.

The Case of the Ogoni

Ogoniland, located on the coastal plains, north of the Niger Delta, in southeast Nigeria, is home to approximately 500,000 people. Shell Oil Company discovered oil there in 1958.[46] By April 1970, Ogoni leaders petitioned the Rivers State government about the hazardous activities of Shell. Over time, the Ogoni began to perceive themselves as "a distinct and unique people," involved in "two grim wars": a "35-year-old ecological war" waged by Shell, and "a political war of tyranny, oppression and greed designed to dispossess the Ogoni people of the rights and their wealth."[47] Consequently, in 1990, the Ogoni issued the Ogoni Bill of Rights in which they articulated demands, including political autonomy within Nigeria and control over the Ogoni environment. Their demands were ignored by both the state and Shell. As a result, the Ogoni internationalized their struggle, appealing to the international community for support.

On January 4, 1993, about 300,000 Ogoni embarked on an unprecedented protest march against the state and Shell Oil Company.[48] The Ogoni explained their grievances in terms of a skewed federal structure, environmental degradation, and destruction of local livelihoods by the oil-related activities of Shell. The success of the protest and the negative publicity generated about Shell's activities in Ogoni became a cause of concern to the oil giant.

In April 1993, Wilbros, an American contractor to Shell, started bulldozing farmlands in Ogoni behind the shield of Nigerian soldiers. The Ogoni protested and the military opened fire on the demonstrators. The situation deteriorated as an Ogoni man was killed, some others sustained severe injuries, and Wilbros's equipments were vandalized. Shocked by the MOSOP attack on Wilbros, and fearing that their clientelist tie to the government and Shell was at risk, 10 traditional elders took a newspaper advertisement wherein they demanded a military crackdown on MOSOP. Apprehensive of negative repercussions for their action, the chiefs fled to Port Harcourt for government protection.

During a meeting of estranged Ogoni elders at the palace of the Gbenemene (King) on May 21, 1994, four prominent pro-government chiefs were murdered in controversial circumstances. The murder provided a pretext for military invasion and occupation of Ogoni. Saro-Wiwa and several activists were arrested, and in November 1995, Saro-Wiwa and eight other Ogoni leaders were executed on trumped-up charges of organizing the murder of the four pro-government chiefs.

The Ogoni Environmental Worldview

A group of warriors and spirit mediums founded Ogoni, and shrines are dedicated to these founding ancestors. Reverence for "the ancestors with their warrior ethos and for the power of the spirit mediums has dominated Ogoni belief."[49] The Ogoni believe in a supreme goddess, Bari, who observes events on earth from the sky, and there are shrines dedicated to the worship of ancestral spirits all over Ogoni. Such belief and ritual practices transform the Ogoni environment into a meaningful place, a "homeland," and a historic territory inhabited by humans and gods, a heritage passed down by the ancestors through the generations, and, therefore, the rightful possession of the Ogoni.

Rivers and streams provide water for human use and seafood for human consumption, but more than that "they are sacred and are bound up intricately with the life of the community, of the entire Ogoni nation."[50] The Ogoni protectively guarded the sanctity of their territory, the independence of their society, and the purity of the environment against external violation. To that end, the Ogoni established a cultural taboo against intermarriage with some neighbors.[51]

At the core of the Ogoni identity is the land, which is the material embodiment of the Ogoni collective memory and the Ogoni prosperity.[52] The Ogoni idea of community evokes images of the connection between the human world and the "other-than-human" world. The land is god and hosts many spirits, which they venerate. The forest, beyond its trees and animals, is sacred, and indiscriminate deforestation is unimaginable.[53] The human and nonhuman animals are coextensive. The essence of a man or a woman could leave its own body and possess an animal.[54] The agricultural planting season in Ogoni is invested with meanings; planting is a social and spiritual occasion.[55] The fruits of the land are honored during festivals.[56] Thus, the Ogoni substantiate Tangwa's claim that within the African metaphysical worldview, the opposition between animals and inanimate things and between the sacred and the profane is unsustainable.

Ogoni Collective Action Frames

A frame refers to an "interpretive schemata that simplifies and condenses the 'world out there' by selectively punctuating and encoding objects, situations, events, experiences and sequences of actions within one's present or past environment."[57] Frames perform three functions: diagnosis of the social condition in need of remedy; prognosis on how to actualize such remedy; and rationale for action. It refers to "the conscious strategic efforts by groups of people to fashion shared understandings of their world and of themselves that legitimate and motivate collective action."[58]

Historically, the Ogoni relied on existing cultural repertoire to overcome existential challenges and, in peace times, maintain their independence.[59] Overwhelmed by a developmental process that divests places of their cultural meaning, the Ogoni sought security and solace in familiar cultural repertoire. Relying on their cultural system and worldview the Ogoni inserted a new emphasis on social justice, arguing that the status quo is unsustainable and inimical to national progress.

Activists framed their struggle in cultural terms as *miideekor,* or the rightful demands of a landlord.[60] In Ogoni, a landlord may lend his field of palm trees to a palm wine tapper. Since culturally the Ogoni have a five-day week, the tapper is expected to tap the trees five days a week. However, he must deliver the product of one day, out of the five days, to the landlord. A day's produce of palm wine meant for the landlord is referred to as *miideekor.* In effect, the Ogoni argued that what they demand is simply their *miideekor,* not everything, from the state and Shell. The latter could appropriate the remaining four day's produce for themselves. Through the *miideekor* frame the Ogoni underlined their ownership of Ogoni. The sense of attachment and ownership is further revealed when Ogoni leader Edward Kobani led a protest to an oil facility to symbolically reclaim Ogoni land. Addressing the crowd later, he asked rhetorically whether the thief did not run away when the rightful owners appeared to take over their patrimony.[61]

The Ogoni leaders developed a common worldview through the frames of divine commission. A moral incentive deployed by MOSOP includes an appeal to the support and involvement of the divine order in their struggle.

Thus, Kobani exhorted Ogoni protesters during the momentous January 4, 1993 protest: "Do not be afraid. Nothing will happen to you on this land which God gave us!"[62] Against all odds, the people believed nothing could hurt them in Ogoni and that their cause would triumph.

Religious belief in specific divine commission can lead to the emergence of leaders who instigate collective action. Saro-Wiwa explains why he walked away from his prosperous literary and business career thus:

> One night in late 1989, as I sat in my study working on a new book, I received a call to put myself, my abilities, my resources, so carefully nurtured over the years, at the feet of the Ogoni people and similar dispossessed, dispirited and disappearing peoples in Nigeria and elsewhere.[63]

The activist linked his leadership role with divine will and truth. The divine purpose was to redeem the Ogoni and other dispossessed groups from their oppression. Thus, Saro-Wiwa frames a divine divide between the otherworldly and the Nigerian polity. In responding to the divine call, Saro-Wiwa accents that divide, siding with the divine.

The political opportunity approach focuses on how political configurations in the polity enhance or foreclose the emergence of social movements. As such, it neglects how religious beliefs and convictions can override objective situations in determining when political authorities could be challenged. The church defined the Ogoni situation as ungodly and worth changing. It likened it to the bondage of Israel in Egypt and emphasized that it was God's will to deliver the Ogoni just as he delivered Israel. The participation of church leaders fostered a spiritually charged context where different religious beliefs converged. In such a syncretistic setting, a mélange of spiritual forces, including the Christian God, Ogoni deities and other actors, fostered the belief in the miraculous.[64]

In the context of a sacred place, activists' emotions intensify, enabling them to dream impossible dreams.[65] Sacred transcendence alters the logic of action, providing a stretched sense of the possible.[66] Reflective of such a sense is the belief in miracles and the ultimate triumph of a just cause regardless of the odds.[67] Activists evinced the belief that their cause would triumph and that nothing could happen to them. Remarking on the pre-protest church service and sermon by Bishop Poromon, Saro-Wiwa claims that after the reading of the Bible passage, an appropriation of Prophet Jeremiah's Lamentation, "I could feel a slight tremor pass through the pews."[68] Continuing, he asserts, "By the end of the service, I found myself greatly encouraged and imbued with a new energy and faith."[69] The beliefs and emotions unleashed through the sacralization of Ogoni and the struggle fostered an expanded sense of the possible.

Religious views altered the sense of the possible, and made it impossible for the Ogoni to reckon with the ruthlessness of the military regimes they confronted. Believing in the ultimate triumph of their cause, MOSOP ignored the authoritarian conditions under which it operated.[70] The Ogoni were emboldened by the belief that their deities, ancestors, and God fully supported and led their struggle. Notwithstanding the constricted and highly repressive political space, the Ogoni proceeded with mobilization against the state. Religious beliefs energized Ogoni conviction that their struggle would triumph regardless of objective power relations. Against that backdrop, the force of authoritarianism waned in comparison to religious convictions.

Interest in the study of frames in the social movement literature has centered on their instrumental and interpretive values.[71] That the past can be transformed into a critical symbolic resource for contention has received little traction.[72] A good example is the use of historical figures for mobilization. The Ogoni movement used Paul Birabi, a foremost Ogoni nationalist leader, as a historical figure. They framed Birabi as a present ancestor and vanguard of the Ogoni struggles. Activists believed he would support and ensure the success of the struggle with every available otherworldly resource. Activists transformed the dead and brought the otherworldly into the mundane affairs of the Ogoni when they made invocative prayers at Birabi's graveyard on January 3, 1993. Belief in the spiritual presence and support of the historical figure deepened assurance that the Ogoni would triumph over their traducers.

Conclusion

In Western environmental scholarship, environmentalism is either about concern for nature or about economic survival. Some argue that major societal transformations have led to a post-materialist Western industrial society, which is less concerned about bread and butter. In that society, environmental concern and interest revolves round the well-being of the nonhuman world. Advocates push the argument further to emphasize that non-Western societies remain materialist as they continue to battle with the challenge of economic survival. Environmental concern in such societies is intrinsically about its survival value. That way, scholars sustain a duality between the material and post-material dimensions of environmentalism. Such dualism reflects in the separation between the this-worldly and the otherworldly. For instance, Western worldview does not attribute spiritual substance to the environment. Rather, concern for nature is shaped merely by the need for harmonious coexistence between human beings and nature, especially in view of the latter's inherent value.

These understandings negate African environmentalism. Africa's environmental values, meanings, and behaviors are not streamlined along economic and this-worldly dimensions. To the contrary, the environment is the material extension of the society. As such, there is little demarcation between this world and the spirit world. Beyond its material and economic value, the environment is the abode of spirits, deities, and ancestors. More than a natural resource base, the environment is invested and shot through with meanings. As the case of the Ogoni shows, concern for the well-being of nature as well as concern for the economic utility of nature are entangled in African environmentalism.

However, evidence for this chapter indicates that the inclusive African attitude to the environment represents only a subset of Nigerian environmentalism. Alongside such an attitude are environmental orientations such as conservationism, which are closer to the Western worldview. Representations of modern environmentalism in Nigeria remain incomplete and misleading when they ignore the various constituting trends and present a part to represent the whole.

Notes

1. Mamdani, "Democratic Theory," 230.
2. Beinart, "African History and Environmental History," 269–302.
3. Madujibeya, "Oil and Nigeria's Economic Development," 305.
4. Ogri, "The Nigerian Petroleum Industry," 11.
5. Buttel and Taylor, "Environmental Sociology," 1992.
6. Ogungbemi, "African Perspective on the Environmental Crisis," 208.
7. Tangwa, "Some African Reflections," 387–95.
8. Tangwa, 389.
9. Tangwa.
10. Tangwa, 389.
11. Ojomo, "Environmental Ethics," 101–13.
12. Guha and Martinez-Alier, *Varieties of Environmentalisms.*
13. Doherty and Doyle, "Beyond Borders."
14. Reno, "Shadow States," 50–51.
15. Fraser, "Rethinking Recognition."
16. Swanson, "Recognition and Redistribution."
17. Walby, "From Community to Coalition."

18. Honneth, "Recognition or Redistribution."
19. Adeola, "Environmental Contamination."
20. Lanternari, "Revolution," 153.
21. Hobsbawm, *Primitive Rebels.*
22. Chakrabathy, *Provincialising Europe,* 16.
23. Chakrabathy, *Provincialising,* 16.
24. Berry, "Concentration without Privatisation?," 66.
25. Moore, "Contesting Terrain."
26. Mehretu, *Regional Disparity in Sub-Saharan Africa,* 77.
27. Mehretu, *Regional Disparity,* 78.
28. Dibua, *Crisis of Development,* 232–33.
29. Beinart, "African History."
30. Madujibeya, "Oil and Nigeria's Economic Development."
31. Dibua, *Crisis,* 215.
32. Beinart, "African History."
33. Adeola, "Environmental Injustice," 44–45.
34. Naanen, "Oil-Producing Minorities."
35. Obi, "Globalisation and Local Resistance," 140.
36. Adeola, "Environmental Injustice."
37. Unerman and O'Dwyer, "The Business Case for Regulation of Corporate Social Responsibility and Accountability," 332–53.
38. Adeola, "Environmental Injustice."
39. Kempton, Holland, Hannan, and Payne, "Local Environmental Groups."
40. Freudenberg, *Not in our Backyards!.*
41. Adeola, "Environmental Injustice," 622.
42. Ibid.
43. Nigerian Conservation Foundation, http://www.ncfnigeria.org/about.php
44. Douglas, "Nigeria: Using a Variety of Advocacy Tools in the Niger Delta."
45. Douglas, "Nigeria."
46. Saro-Wiwa, *Genocide,* 65.
47. Saro-Wiwa, *A Month,* 148.
48. MOSOP, *Ken Saro-Wiwa's Last Words.*
49. Kpone-Tonwe and Salmon, "The Arts of the Ogoni," 276.
50. Ibid.
51. Kpone-Tonwe, 2006 interview, Port-Harcourt.
52. Livesey, "Royal/Dutch Shell," 73.
53. Saro-Wiwa, *Genocide,* 12.
54. Saro-Wiwa, 12
55. Ibid.
56. Ibid.
57. Snow and Benford, "Master Frames," 137.
58. McAdam, McCarthy, and Zald, "Introduction," 6.
59. Kpone-Tonwe, 2008 interview, Port-Harcourt.
60. Agbonifo, interface.
61. Saro-Wiwa, *A Month and a Day.*

62. Saro-Wiwa, *A Month,* 121.
63. Ibid.
64. Agbonifo, "Territorialising Niger Delta Conflict."
65. William H. Sewell Jr, "Space in Contentious Politics."
66. Aminzade and Perry, "The Sacred, Religious and Secular."
67. Tschirgi, *Turning Point.*
68. Saro-Wiwa, *A Month,* 121.
69. Saro-Wiwa, *A Month.*
70. Agbonifo, "Marginalised Violent Internal Conflict."
71. Jansen, "Resurrection and Appropriation."
72. Jansen, "Resurrection and Appropriation," 958.

Bibliography

Adeola, F.O. "Environmental Contamination, Public Hygiene, and Human Health Concerns in the Third World: The Case of Nigerian Environmentalism." *Environment and Behaviour* 28 (1996): 614–46.

Adeola, F.O. "Environmental Injustice and Human Rights Abuse: The States, MNCs, and Repression of Minority Groups in the World System." *Human Ecology Review* 8 (2001): 39–58.

Agbonifo, J. "The Marginalised Violent Internal Conflict (MVIC) Model and the Ogoni Conflict in the Niger Delta, Nigeria." MA diss., American University in Cairo, 2003.

Agbonifo, J. "Territorialising Niger Delta Conflicts: Place and Contentious Mobilisation." *Interface* 3 (2011): 240–56.

Beinart, W. "African History and Environmental History." *African Affairs* 99 (2000): 269–302.

Berry, S. "Concentration without Privatization? Some Consequences of Changing Patterns of Rural Land Control in Africa." In *Land and Society in Contemporary Africa,* edited by S. Reyna and R. Downs, 53–75. Hanover, NH: University Press of New England, 1988.

Chakrabathy, D. *Provincializing Europe, Postcolonial Thought and Historical Difference.* Princeton: Princeton University Press, 2000.

Dibua, J.I. *Modernization and the Crisis of Development in Africa: The Nigerian Experience.* Aldershot: Ashgate, 2006.

Douglas, O. "Nigeria: Using a Variety of Advocacy Tools in the Niger Delta." www.eli.org/pdf/advocacytoolscasestudies/casestudy.nigeria.final.pdf

Fraser, N. "Rethinking Recognition." *New Left Review* 3 (2000): 107–20.

Freudenberg, N. *Not in Our Backyards! Community Action for Health and the Environment.* New York: Monthly Review Press, 1984.

Guha, R., and J. Martinez-Alier. *Varieties of Environmentalisms: Essays North and South.* London: Earthscan, 1997.

Hobsbawm, E.J. *Primitive Rebels.* Manchester: University of Manchester, 1959.

Honneth, A. "Recognition or Redistribution?: Changing Perspectives on the Moral Order of Society." *Theory, Culture and Society* 18 (2001): 43–55.

Jansen, R.S. "Resurrection and Appropriation: Reputational Trajectories, Memory Work, and the Political Use of Historical Figures." *American Journal of Sociology* 112 (2007): 953–1007.

Kempton, Willett, Dorothy C. Holland, Katherine Bunting-Howarth, Erin Hannan, and Christopher Payne. "Local Environmental Groups: A Systematic Enumeration in Two Geographical Areas," *Rural Sociology* 66(4) (2001): 557–578.

Kpone-Tonwe, S., and J. Salmons. "The Arts of the Ogoni." In *Ways of the Rivers: Arts and Environment of the Niger Delta,* edited by M.G. Anderson and P.M. Peek, 275–301. Los Angeles: UCLA Press, 2002.

Lanternari, V. "Revolution and/or Integration in African Socio-Religious Movements." In *Religion, Rebellion, Revolution,* edited by Bruce Lincoln, 117–41. New York: St. Martin's Press, 1985.

Livesey, S.M. "Eco-identity as Discursive Struggle: Royal/Dutch Shell, Brent Spar, and Nigeria." *Journal of Business Communication* 38 (2001): 58–91.

Madujibeya, S.A. "Oil and Nigeria's Economic Development." *African Affairs* 75 (1976): 284–316.

Maier, K. *This House Has Fallen: Nigeria in Crisis.* London: Penguin Books, 2000.

Mamdani, M. "Democratic Theory and Democratic Struggles (in Africa)." In *Government and Politics in Africa—A Reader,* edited by O. Nnoli. Harare: African Association of Political Science (AAPS) Books, 2000.

McAdam, D., J.D. McCarthy, and M.N. Zald. "Introduction." In *Comparative Perspectives on Social Movements: Political Opportunities, Mobilizing Structures, and Cultural Framings,* edited by Doug McAdam, John D. McCarthy, and Mayer N. Zald, 1–20. Cambridge: Cambridge University Press, 1996.

Mehretu, A. *Regional Disparity in Sub-Saharan Africa: Structural Readjustment of Uneven Development.* Boulder, CO: Westview Press, 1989.

Moore, D.S. "Contesting Terrain in Zimbabwe's Eastern Highlands: Political Ecology, Ethnography, and Peasant Resource Struggles." *Economic Geography* 69 (1993): 380–401.

MOSOP. *Ken Saro-Wiwa's Last Words.* Port-Harcourt: MOSOP, 2004.

Naanen, B. "Oil-Producing Minorities and the Restructuring of Nigerian Federalism: The Case of the Ogoni People." *Journal of Commonwealth and Comparative Politics* 33 (1995): 46–78.

Obi, Cyril I. "Globalisation and Local Resistance: The Case of the Ogoni versus Shell." *New Political Economy* 2 (1997): 137–48.

Ogri, O.R. "A Review of the Nigerian Petroleum Industry and the Associated Environmental Problems." *The Environmentalist* 21 (2001): 11–21.

Ojomo, P.A. "Environmental Ethics: An African Understanding." *The Journal of Pan African Studies* 4 (2011): 101–13.

Reno, W. "The Politics of Insurgency in Collapsing States." *Development and Change* 33 (2002): 837–58.

Reno, W. "The Politics of Violent Opposition in Collapsing States." *Government and Opposition* 40 (2005): 123–327.

Reno, W. "Shadow States and the Political Economy of Civil War." In *Greed and Grievances: Economic Agendas in Civil Wars,* edited by Mats Berdal and David M. Malone, 43–68. Boulder, CO: Lynne Rienner, 2000.

Saro-Wiwa, K. *Genocide in Nigeria: The Ogoni Tragedy.* Port-Harcourt: Saros International Publishers, 1992.

Saro-Wiwa, K. *A Month and a Day: A Detention Diary.* Ibadan: Spectrum Books Limited, 1995.

Sewell, W.A. Jr. "Space in Contentious Politics." In *Silence and Voice in the Study of Contentious Politics,* edited by R.R. Aminzade, J.A. Goldstone, D. McAdam, E.J. Perry, W.H. Sewell Jr, S. Tarrow and C. Tilly, 51–88. Cambridge: Cambridge University Press, 2001.

Snow, D.A. and R. Benford. "Master Frames and Cycles of Protest." In *Frontiers of Social Movement Theory,* edited by Aldon D. Morris and Carol McClurg Mueller, 133–55. New Haven, CT: Yale University Press, 1992.

Stolper, W.F. *Planning without Facts: Lessons in Resource Allocation from Nigeria's Development.* Cambridge, MA: Harvard University Press, 1966.

Swanson, J. "Recognition and Redistribution: Rethinking Culture and the Economic." *Theory, Culture and Society* 22 (2005): 87–118.

Tangwa, G. "Some African Reflections on Biomedical and Environmental Ethics." In *A Companion to African Philosophy,* edited by Kwasi Wiredu, 387–95. Oxford: Blackwell Publishers, 2004.

Tschirgi, D. *Turning Point: The Arab World's Marginalization and International Security after 9/11.* Westport: Praeger Security International, 2007.

Unerman, Jeffrey and Brendan O'Dwyer. "The Business Case for Regulation of Corporate Social Responsibility and Accountability." Accounting Forum 31 (2007): 332–53.

Walby, S. "From Community to Coalition: The Politics of Recognition as the Handmaiden of the Politics of Equality in an Era of Globalization." *Theory, Culture and Society* 18 (2001): 113–35.

World Bank. *Nigeria: Options for Long-Term Development.* Baltimore: The Johns Hopkins University Press, 1974.

Chapter 10

Iraqi Hybrid Environmentalisms: The Local-Global Nexus?

M. A. Mohamed Salih

Introduction

Environmental consciousness is present in all societies whether ruled by democratic or authoritarian states. However, the manifestations of this consciousness in terms of environmental movements differ from one type of political regime to another. While democratic regimes allow environmental consciousness to be channeled and expressed through freely formed political, social, and environmental movements, the formation and activities of such environmental organizations and movements are brutally suppressed under authoritarian regimes.

Saddam Hussein's Iraq belongs to authoritarian regimes in which the emergence of freely organized environmental movements was inconceivable under a one-party state where the Ba'ath Party controls all aspects of human affairs, imposing itself as the only political organization in the country. The absence of conventional environmental movements and a green public in Saddam's Iraq is understandable. However, this should not mean that there were no underground environmental movements and a green public disguised under various forms of mainstream or political opposition. As has already been clearly demonstrated in other cases, Iraq's underground green movement took the form of self-determination movements in the areas inhabited by the Marsh Arabs, or movements against the occupation both nationally and internationally.

The Iraq wars brought about their own environmental problems and its aftermath unleashed nascent environmental movements nationally that gained considerable support from global antiwar movements. This chapter elucidates the nexus between local and global environmental movements in the post–Saddam Hussein regime, defying the conventional Western liberal conceptions of single-issue post-materialist environmental movements, by arguing that Iraq's national environmental movements and global environmentalists such as "No Oil No War" have one common characteristic—that is, both are hybrid environmental movements, using multiple identities: environment, self-determination, deterioration of livelihoods, social justice, and so forth.

Iraq suffers from internally and externally induced environmental crises such as the neglect of water and sanitation structures, land degradation and depletion of the urban infrastructure due to the sanctions regime and the environmental impacts of the Gulf War. There is also the menacing presence of the impaired environment of the Arab marshlands, which were drained during the Saddam regime in order to control the Shia opposition. Iraq's environmental crises (pollution by oil production and military industry controlled by global networked markets and interests and the transboundary nature of the Marsh Arabs' lands) are both local and global, thus blurring the boundaries between local and global environmentalism. Both environmentalisms are hybrid and far from being single-movement issues.

During Saddam's authoritarian rule, environmental struggle was by-and-large considered a metaphor for political opposition and the right to self-determination by other means, epitomized by the underground environmentalism of the Marsh Arabs. The antiwar activists campaigned in alliance with environmentalism under the "No War, No Warming" movement. In post-Saddam Iraq, the interests of the global and local environmental movements diverged: globally, the relationship between oil or fossil fuel consumption and climate change took center stage. Locally, rehabilitating the destroyed urban infrastructure and the more than 300 hotspots of environmental hazards, addressing the issues of the Marsh Arabs, and the return of the displaced peoples were major concerns.

To be sure, Iraqi environmentalism and the international antiwar-cum-environmental movements could be viewed partly as an extension of the struggle against the Iraq War being really an instrument for controlling Iraq oil, which would contribute further to global warming, and partly locally as a struggle to reverse the environmental damage caused by war, industrial pollution, and the draining of the Marsh Arabs' marshes. Globally, it united the antiwar and global environmentalism, using the Iraq War as a platform to revive the stalled debate on the Kyoto Protocol following the U.S. failure to ratify in 2000. Iraqi local and global environmentalisms are represented by different interests within Iraqi and global environmental publics.[1]

This chapter consists of four sections: The first section is a background that describes and explains four major types of environmental degradation in Iraq. It introduces spillage caused by the oil industry and the Gulf War, and the environmental crisis of the Arabs' marshes. These environmental crises are delineated as major factors contributing to the rise of hybrid national and global environmental movements. The second section is concerned with Iraqi hybrid environmentalism, that is, environmental movements whose identities traverse the boundaries of two or more movements and are both environmental and social. The third section introduces the hybrid nature of the global environmental movements that articulated an antiwar stand linked with global social environmental issues such as climate change. The fourth section attempts to argue that local and global environmental movements are multimovements, linking the antiwar campaigners with Iraqi emergent green publics, parties, and movements.

Background

This brief background describes and explains four major types of environmental degradation in Iraq: spillage from the oil industry and the Gulf War, environmental degradation, and the crisis of the Marsh Arabs' marshes. It is essential in order to understand environmental degradation as a factor in the rise of environmental consciousness that contributed to the emergence of local environmental movements and a politically active green public, including the National Green Party of Iraq (NGPI). Antiwar and environmental movements within Iraq were supported by the global antiwar movement, which enjoyed considerable support from the global environmental movement. I take these four sources of Iraq's current environmental crisis in turn.

Before fully recovering from the impact of the Iraq War (1980–1988), Saddam orchestrated two wars within little more than a decade (in 1991 and 2003), which brought their own environmental devastation. A UNEP study on the impacts of the 2003 war on Iraq reported that:

> Iraq's industry has suffered from a decade of economic sanctions and lack of investment. This has led to chronic environmental problems, such as discharges of untreated effluent to surface waters, spillages and discharges of chemicals to soils and groundwater, and widespread uncontrolled emission

> of particulates and gases from stacks. The Iraq War has undoubtedly exacerbated the chronic environmental stresses that have accumulated in Iraq over the past two decades. An important part of the environmental damage associated directly with the war arises from the looting and pillaging of key infrastructures and the ransacking of equipment and supplies, including the looting of hazardous and radioactive materials.[2]

Second, industrial pollution has worsened after the war as the industrial sector, including the oil industry, was subjected to economic sanctions[3] which caused chronic environmental problems such as the discharge of untreated wastewater from factories into rivers, soil and groundwater contamination, and air pollution as a result of releasing poisonous gases and particulate matter into the atmosphere.[4]

Third, there is the destruction of urban environmental infrastructure and widespread pollution and waste. The Iraqi ministry of environment outlined the major environmental impacts of war as directly felt by the people in their daily lives as follows: there is a shortage of safe drinking water; there is no electricity (as most Iraqis complain); poor quality of sanitation and sewage and wastewater disposal; there are sheep and cattle grazing between houses and near schools; and space between houses and schoolyards is filled with garbage and stagnant water due to the destruction of the sewage system. The way the Iraqis experienced the destruction of their urban infrastructure during the war is not just the sentimental view of an angry lot. Studies conducted by social and environmental scientists reveal that the incidence of such problems concords with what Iraqi people have seen vividly in the surrounding environment.[5] Munitions extracted from depleted uranium molded into bullets and bombs used as armor-penetrating weapon designed to destroy tanks, armored personnel carriers, released carcinogenic dust, as did bunkers. When stuck in the lungs, it wreaked havoc in the form of tumors, hemorrhages, ravaged immune systems, and leukemia. This explains why the incidence of leukemia in Iraq has increased over 600 percent. Detection and treatment of cancers was made unnecessarily difficult by Iraq's forced isolation under a regime of sanctions, which had produced a humanitarian crisis of its own.[6]

Fourth, the draining of the Southern marshes after the Marsh Arabs' revolt against Saddam in the wake of the Iraq War in 1991 caused considerable damage to both the environment and sources of subsistence for the inhabitants of the marshes. As a policy intended to subdue rebellion and preventing them from using the marshes as a military base of insurgency operations, the government drain and transform them into farm lands.

According to UNEP:

> by 1970 the wetlands covered nearly eleven thousand square kilometers; after the draining they covered fewer than a thousand. It is estimated that over five hundred thousand Marsh Arabs were forced to take refuge in other parts of Iraq.[7]

As will be elaborated in the following sections of this chapter, the post-Saddam environmental crisis has contributed to the emergence of two types of hybrid environmental social movements: A divided internal movement consisting of first, a pro-Saddam insurgency that resisted the occupation as a major factor in the deterioration of the environmental and livelihood conditions; and second, anti-Saddam (pro-occupation) movements such as the Marsh Arabs, whose livelihood suffered as a result of the draining of the southern marshes. Globally, a hybrid environmentalism (No War No Warming)[8] saw the war as really being a war about oil, and resisted it on the grounds that fossil fuel is a major contributor to global warming and subsequently climate change.

The war-related environmental damage caused by the Iraq war has spread well beyond Iraq as reported by the U.S. and British military establishment.[9] Some of the most gruesome hallmarks of this war are death or injury and psychological problems associated with war trauma and the so-called Gulf War syndrome. From this perspective, the Iraq war is no longer a distant event whose consequences are watched with any sense of neutrality. In a sense, the ecological footsteps of the Gulf War syndrome are global because it has affected those who participated in it whether they are Iraqi, American, British, or fighters and citizens of any other nationality that participated in the war effort. Obviously, the nature of modern warfare, which is globally networked, is similar to environmentalism, which is equally globally networked.

At least four conclusions can be made:

1. Environmental problems and the respective emergent environmentalism in post-Saddam Iraq cannot be separated from the war that produced its own environmental hazards as well as exacerbated an already desperate situation.
2. The movements for a clean and healthy environment are of at least two types: urban-based, which were at times exploited by the insurgency and concerned mainly with oil spills, destroyed chemical and biochemical industries and urban infrastructure; and land-based, such as the Marsh Arabs' movement,

which combined environmental and fundamental social problems (poverty, ill-health, malnutrition, high levels of illiteracy), and which cannot be treated in isolation from their struggle for self-determination, which would enable them to use environmental resources denied to them by an authoritarian regime.

3. In the West, the opposition to the Iraq war has united the antiwar and environment movements, with the latter advocating the view that an antiwar stance would prevent Western oil interests from controlling Iraq oil and by doing so, reducing carbon contribution to global warming.
4. Iraq's internal and external environmentalisms are hybrid in the sense that they combine more than one issue because the anti-war protest was championed by environmentalists as well as women and civil rights and labor unions and green political parties in the West.

In the following sections, I explain the factors contributing to the hybridity of Iraqi environmentalisms and their relationship with the global environmental movement that found resonance in the "No War No Oil" movement.

Iraqi Hybrid Environmentalisms

Hybrid environmentalism refers to environmental movements that combine environmental issues (land, water, forests, climate change, biodiversity) and one or more social movements (women, civil rights, peace, peasant, or social justice). Rojas and Heaney articulate this as follows: "Organizations with hybrid identities possess three advantages that facilitate their use of interorganizational networks in a way that promotes spillover between movements. These advantages include, first, the ability to satisfy the individual identity needs of activists; second, the capacity to serve the organizational maintenance needs of social movements; and third, an enhanced likelihood to achieve strategic positions within interorganizational networks."[10] As the case of the Marsh Arabs illustrates, the environment or the expanses of water resources on which they live and work form a central part of their struggle for self-determination or liberation from Saddam's Ba'athist regime (1979–2003) that ruled Iraq with an iron fist until its demise in the second Iraq war in 2003. Below I briefly introduce the Arab Marshes and the Marsh Arabs and their hybrid rebellion.

Iraq's Mesopotamian marshes are known as the Arab Marshes and also popularly referred to as the "Garden of Eden." *Mesopotamia* means "between rivers," referring to the location between the Tigris and the Euphrates. Geographically, the heartland of the marshes comprised three principal areas: a) the al-Hammar Marshes, located south of the Euphrates between al-Nasiriyya and Basra; b) the Central Marshes, located between the Tigris and the Euphrates rivers in a triangular area bounded by al-Nasiriyya, al-Qurna, and Qal'at Salih, with a section further north around the city of al-'Amara (commonly known as the al-'Amara Marshes); and c) the al-Huwaizah Marshes, located east of the Tigris and extending into Iran (where they are known as the al-Azim Marshes). Together, these wetlands formed a series of interconnected permanent marshes and lakes covering an area of some 8,800 square kilometers, extending to some 20,000 kilometers when large tracts of dry or desert land were seasonally inundated.[11]

The Arab Marshes were once the largest wetlands in southwest Asia and covered more than 15,000 square kilometers. Since the 1950s, the basin's ecology has been fundamentally altered by the construction of dams and drainage schemes (including major projects in Turkey, where the Tigris rises), which have substantially reduced the water supply and eliminated the flood pulses that have historically sustained the wetland ecosystems in the lower basin. Following the end of the Iran-Iraq War, the diking and diversion of Euphrates waters into the canal originally intended solely for saline irrigation waters greatly accelerated the drying out of the Al Hammar marshes to the northwest of Al Basrah. A large swathe of the Central Marshes was drained to facilitate the movement of military units during the Iran-Iraq War, and in 1992 a major drainage project was initiated in this region.[12] After the draining of the Marshes by Saddam Hussein, the only remaining marsh was the northern portion of Al-Hawizeh, which straddles the border between Iran and Iraq. The other two marshes, Al Qurna and Al-Hamma, were virtually destroyed by 2000. The remaining Al-Hawizeh was only 35 percent of its 1977 size of 3,076 square kilometers by 2000.[13] According to UNEP (2001), both the Al Hammar and the Central Marshes no longer exist. By 1994, following the canalization of the two main Tigris tributaries replenishing the Al Hawizeh marshes, these were drying out quite rapidly, which coincided with the construction of Ataturk Dam in Turkey, which caused extensive areas of the marshlands to disappear completely to be replaced by bare land and salt crusts that are incapable of supporting traditional Marsh Arab society, the existence of which has, from time immemorial, been entirely dependent on the wetland ecosystem. In 1973 the marshlands covered an area of 8,926 square kilometers, extending to 20,000 square kilometers during seasonal inundation of land. By the year 2000 this had been reduced to just 1,297 square kilometers.[14]

The Marsh Arab studies are rare and most of those that exist use old sources[15] to infer about their present interactions with the rest of Iraqi society and the central political institutions. Marsh Arab studies describe them as a confederation of Muslim Shia tribes that inhabit the southeastern Iraqi borders with Iran. They consist of a number of semiautonomous tribes such as Albu Muhammad, Ferayghat, Shaghanbah, and Bani Lam, the Ma'dan.[16] They have a unique culture within Iraq, which earned them prejudice from mainstream Iraqi Arabs—of mixed Arab, Persian and other non-Arab origin, they speak a distinct Arabic dialect, and are suspected to harbor loyalty to Iran due to geographic proximity and the affiliation of the majority with the neighboring Shia population of Iran.[17]

The marshland environment sustains the production of rice, wheat, barley, millet, and rice as well as livestock husbandry (cattle, sheep, and buffalo) and fishing. The draining of the marshes has exerted tremendous pressure on peoples' livelihood and land-based economy. This situation has been aggravated by land acquisition by Saddam's government for large-scale irrigation schemes and private concessions to Sunni Iraqis.[18]

The struggles of the population of the Marsh Arabs can be viewed as an environmental struggle both against land alienation and pollution by the oil industry and against the draining of the marshlands. Their struggle is also a liberation for self-determination against an authoritarian regime and their stigmatization by Saddam's regime, which treated them as second-class citizens thus bypassing their basic needs—health, education, clean drinking water, employment, infrastructure, and other social amenities. This was captured succinctly by Juan Cole, who described it as follows:

> Ba'athist social engineering of the Marsh Arab areas began in the 1980s. Some of the impetus came from the Iran-Iraq War, which spurred the regime to conscript many southern Shiites, and to engage in extensive building of defensive facilities. For instance, the al-Ma'rada tribe inhabited the wetlands around the Majnun oil fields, the target of a major set of Iranian offensives. The Ba'ath regime built enormous earthworks to keep the Iranians at bay.[19]

Hence, the 1991 rebellion of the Marsh Arabs (also known as the Shia rebellion) against the Saddam regime was an extension of a process that began shortly after Saddam rose to power in 1979. The Saddam regime's attempt to destroy the Marsh Arabs' sources of living in order to break their militancy against his rule was not a simple punishment for their support of the invading coalition forces, but dates back to more than a decade long mass displacement, marginalization, and land alienation.[20] For example, during the 1970s, the Iraqi government embarked on developing expansive agricultural projects, using large quantities of water from the Tigris and in 1985, draining part of the marshes for oil exploration.[21] These activities have already drained part of the marshes to a noticeable level.[22]

The 1991 operations that drained the marshlands were unrelenting, aggressive, and cynically inhumane. They were unrelenting and aggressive because by the end of 2000, according to United Nations Environmental Program (UNEP) estimates, 90 percent of the marshlands had disappeared and with it the sources of living of over 2 million people. They were cynically inhumane because the drainage canals were given names that glorified this murderous intent, such as *Al-Amarah* (glory) and *Al-Izdihar* (prosperity). Far from glory and prosperity, the draining of the marshes brought ruin and suffering to the inhabitants of the Arab marshlands. Sadly, it is estimated by Nicholson[23] that more than 500,000 Ma'dan have been displaced, 95,000 of them to Iran, 300,000 are internally displaced, and the remainder became refugees in other countries.[24]

From this perspective, the draining of the Arab marshlands represents the culmination of the Ba'ath policy to control the marshes, thus controlling people by engineering and controlling the environment. The Ma'dan guerrillas' attacks against the Ba'ath regime, before and after the second Iraq War and the draining of the marshes,[25] combined both environmental and liberation or self-determination struggles, thus making the distinction between the two superfluous. I argue that elsewhere in authoritarian regimes, the environment is a determining factor in the rise of liberation movements and that these movements are not merely "ethnic," "tribal," clan-based or class-based. These liberation struggles aiming at reclaiming the state are also movements against state interventions that resulted in depleted environmental resources and loss of sources of livelihood.[26]

The liberation struggles in the Marsh Arabs cannot be treated separately from the environmental conditions within which politically oppressed, impoverished rural peoples engaged, as they cannot afford to distinguish between issues of human rights, self-determination, environmental justice, and dwindling survival possibilities. In essence, compartmentalizing liberation struggles into political, environmental, social and so forth is no longer possible as in many cases, these movements intersect and in many others, acquire multimovement identities. This is particularly so in authoritarian regimes where the absence of democracy (political or environmental) gives way to state violence against both people and nature. Those displaced or who experienced environmental alienation use counterviolence to protect their environmental life support system. The struggles the inhabitants of the Arab marshlands waged against the Ba'ath regime following the environmental crises inflicted on them by the Saddam regime are both liberation and environmental struggles and bear the form and content of hybrid environmentalism.

Attempts at reversing the environmental and livelihood destruction caused by the draining of the Arab marshlands began in 2003, immediately following the occupation of Iraq by the allied forces. Iraq's emergent environmentalism has also been shaped by the harsh realities created by the insurgency and the steady deterioration of the living conditions of the majority of the population. The economic sanctions and war have reduced Iraq from a second world economy to one of the poorest in the Middle East.[27] The insurgency, both local and foreign, destabilized large swathes of the country, with the majority of the population confined to displacement camps or large, insecure, and highly militarized urban centers. However, the situation in southern Iraq was relatively stable after degenerating into running battles between the Iraqi insurgents and the occupation forces led by Britain. Unlike the Iraqi Sunni-dominated areas, southern Iraqi life returned to a semblance of normality and some of the displaced Marsh Arabs returned, with the hope of reconstructing their shattered lives. In 2004, assisted by the Iraqi ministry of water resources, the local population began re-flooding about 40 percent of the marshes, but because of the haphazard way in which this re-flooding was carried out, the rehabilitation of the marshes has been patchy, with some localities showing rapid restoration of vegetation while others have been slower to recover and still others showing no signs of recovery whatsoever. By the end of 2004, over 40,000 displaced inhabitants returned and resettled the marshes, facing many problems related to public hygiene and quality of life owing to the loss of the fisheries and other elements that formerly provided sustenance.[28]

The post-Saddam era witnessed the emergence of several environmental movements and NGOs. It is difficult to give an impression of a thriving environmentalism, but it is safe to argue that hybrid movements organized around major social issues (poverty, health, and food) often find themselves confronted with major environmental problems such as water and soil contamination from chemicals, oil spills, and open sewage. The Eden Again Project for the restoration of the Arab Marshes is one such example.

The Iraq Foundation's Eden Again Project was initiated with the aim of supporting efforts to restore the marshlands. In June 2002, the project team began to assemble an international panel of scientists to provide advice on the technical aspects of marshland restoration and inform the development of technical planning documents that could lay the foundation for such restoration (the International Technical Advisory Panel).[29] In 2003, the Iraqi people began to restore water flow to the marshes, and the Eden Again Project moved to Iraq to help them make restoration a reality.

The Eden Again project involved multiple actors at the international, regional, and local levels. Internationally, Eden Again gauged the support of the United Nations Sustainable Development Commission (UNSDC), the United Nations Environmental Program (UNEP), as well as the support of bilateral development cooperation from Japan, Italy, and the United States.[30]

Regionally, the Eden Again project sought the experiences gained by Azraq Reserve of the Jordanian Royal Society for the Conservation of Nature, the only marshland reserve in Jordan. It found the training courses on nature conservation management used in Azraq Reserve relevant for the Iraqi trainees entrusted with the implementation of the project.[31]

Locally, environmental movements working with the Eden Again Project,[32] which include Nature Iraq and the Arab Marshlands Forum, are as much concerned with the environment as with education, humanitarian assistance, and human rights. The reverse is also true. Organizations such as the Iraq Foundation for Democracy and Human Rights have joined hands with the University of Basra to conduct research into water and aquatic organisms, and to survey the recovery status of ecosystems following the re-flooding of the southern marshlands.[33]

The Eden Again project is concerned both with conservation and livelihood issues. It comprised activities such as health clinics, clean water supply, agricultural production and agribusiness, and livestock and dairy production that require the participation of diverse local actors (governmental and nongovernmental as well as regional and international). This multilayered participation in the Eden Again Project illustrates the local-global linkages and the multi-scale level of environmental action to restore the Arab Marshes. It is also a clear case of a divided environmental movement in Iraq where the Eden Again project is largely sponsored by anti-Saddam, local, environmental-cum-rebel movements and the support they have gained from global environmental movements and environment conservation support organizations.

The environmental destruction caused by Saddam's political oppression and the destruction of the urban infrastructure by wars contributed to the emergence of urban-based environmental movements in the post-Saddam era. While a few zealous citizens joined the insurgency or highly politicized sectarian groups, the majority of the Iraqi people resisted responding to violence with violence, and organized themselves in social environmental movements and political parties, including green political parties and movements. These movements and parties are both political and environmental. The possibility of creating single-issue environmental movements was almost impossible for Iraqis, whose capacity to resist is weakened by starvation and shortage of the basic necessities of life.

Little surprise therefore that the Iraqi environmental movement called simultaneously for environmental rehabilitation and the improvement of the living and environmental conditions available for the Iraqi peoples. This has been true whether in the case of the Marsh Arabs in the far South or, clearly so, in the case of Iraq's first green political party, the National Green Party of Iraq (NGPI), which was established in 2003, calling on every Iraqi to plant a date tree as part of its program for greening Iraq. Its motto is: "For a green Iraq free from environmental, structural and intellectual pollution and corruption." As a secular party, the NGPI calls for "a green movement against sectarianism." Among the main objectives of the NGPI are the following: environmental rehabilitation and greening of Iraq, and cleaning the land from chemical contamination, creating a free democratic and united Iraq, based on multiparty democracy, justice and the rule of law, freedom of religion and respect for human and minority rights. The NGPI is an affiliate of the Global Greens and therefore shares its antiwar stance and calls for the peaceful resolution of all conflicts.

In sum, political and social activities outside the Ba'ath Party were banned during Saddam's rule, but his demise meant an environment conducive for the emergence of social, environmental, and political organizations. A national ecology party with international affiliations and offices has been established. Hundreds of environmental movements with international partners and supporters have also been created along class, ethnic, religious, and regional relations. A common feature of these environmental movements is hybridity, combining environmentalism with other major social concerns. A second feature is that although movements such as those of the Marsh Arabs appear local in their concerns, some of the problems associated with the draining of the marshes are global (biodiversity loss and disappearance of transnational migratory birds' breeding grounds), and therefore invite the support of the global environmental movements.

No War No Warming: Hybrid Global Environmentalism

While the struggle of some environmental movements in Iraq such as the Marsh Arabs emerged from within the embryo of war, applied war tactics and armed rebellion against the state, and even supported the invading coalition forces against Saddam's regime, on the contrary, the major green political parties and environmental movements in the West were antiwar. The second form of hybrid environmentalism emerged immediately after the second Iraq war in 2003. This environmentalism has drawn support from the antiwar movement and vice versa. This hybridity is informed by the global context of climate change negotiations, and the Bush administration's decision (in 2001) to pull the United States out of the Kyoto Protocol. President George Bush dismissed the Kyoto Protocol on the grounds that they were too costly, unrealistic, and if ratified, would lead to the loss of millions of jobs. The fossil fuel corporate interests, whose interests the Bush administration represented, were blamed for undermining the Protocol, given the fact that the United States is one of the largest polluters in the world. In other words:

> by the early 2000s, the American environmental movement found itself going through hard times. Numerous observers declared that the environmental movement was "dead," or at least needed to be revamped in order to address the challenges of global warming.[34]

The Iraq War came at a time when the environmental movement was rather demoralized by the United States' and Australia's negative responses to Kyoto. The second Iraq war was described by the position statement against the war published by Global Greens (Global Green Action Proposal April 2003), which made a direct link between the war and the United States' and Britain's interests to control Iraqi oil reserves. The statement reads as follows:

> The preparedness of the United States, Britain and supporting governments to act unilaterally against Iraq, despite the fact that the United Nations weapons inspectors have been accepted by the Iraq government and are still at work, despite the fact that Iraq has not attacked any country and is not a threat to world and regional peace, reveals that the real purpose of such a war is to gain control of the world's second largest oil reserves and to install a regime sympathetic to the economic and political objectives of the United States and Britain. The United States is determined to protect its corporate oil interests and its huge energy consuming society and lifestyle at any cost.

Within the United States, the U.S. Green Party published a statement against the Iraq War arguing that the United States should:

> take substantial steps to free our foreign policy from distortion by U.S. economic dependency on energy from non-renewable fossil fuels. Encourage energy conservation and invest heavily in the development of renewable energy. It should also shift investment from military spending to address the basic needs of humanity, both within the United States and around the globe: clean air and water, access to education and health care, adequate housing and transportation, and civil liberties.[35]

No War No Warming according to Rojas and Heaney, was:

> founded by climate activists who sought to partner with antiwar leaders to better link the Iraq War to global warming. The organization stressed connections among several issues such as that oil is the mutual cause of war and warming, military operations consume enormous quantities of fossil fuels, and war and warming similarly harm the neediest populations of the world.[36]

In 2007, No War No Warming published a short statement to explain its objectives where it argued that burning oil and gas is responsible for more than a third of global greenhouse gas emissions; the military is the largest single consumer of petroleum in the United States—which is, in turn, the largest consumer of petroleum in the world. The oil industry has systematically funded efforts to discredit legitimate climate science and obstruct international negotiations for more than a decade. Big Oil's legislators and lobbyists continue to block meaningful action on clean energy alternatives at every opportunity, just as they continue to fail to stop the war. Big Oil's lobbyists and the Bush administration were pushing the proposed Iraqi Oil Law, to open Iraq's oil fields to be controlled by foreign corporations like Exxon and Chevron.[37]

To be sure, in addition to institutionalized green political parties, which supported No War, No Warming, the campaign was joined by a broad array of single-issue social movements and organizations under the slogan: "Help end the Iraq war and an action against global warming by demanding separation of oil and state."[38] Like their Iraqi counterparts, single-issue movements are increasingly connected with other broader and related issues thus applying the old rule of ecology: "everything is connected with everything else." Environmental and antiwar movement pragmatic coalitions, on the one hand, echo the truism that concerns regarding the environmental life support system are integrative both in function and relationship with each other. On the other hand, they are expressions of environmental social relations operating within an interdependent local-global nexus. In the following section, a brief synthesis of these relations and solidarities and the conflicts they invoke in a reconfigured global environmentalism is provided.

Arguing the Local-Global Nexus

The Iraq case illustrates the hybrid nature of local and global environmentalism owing to the emergence of globalized environmental problems in globally networked environmental social movements. If globalization is about de-territorialization and the annexation of time and space through production processes to which energy production is essential, the global environment is highly integrated and its functions (if only to remind the reader: life support, production, sink and regulatory functions, to mention but a few) are also integrated.

While hybrid environmentalism could from the first instance be seen as a response to local environmental hazards and their direct effect on peoples' livelihoods, at the larger synthesis, these struggles are concerned with global environmental issues that are not part of the environmental consciousness of the Marsh Arabs: global warming and loss of biodiversity. The UNEP 2000 and 2003 reports on the draining of the marshlands have revealed the loss of a migratory and birds' breeding ground:

> birds migrating from Eurasia to Africa, and consequent decrease in bird populations in areas such as Ukraine and the Caucasus and extinction of several plant and animal species important for human uses in the form of pharmacological and other health usages.[39]

Likewise, Western environmentalism is hybrid and both local and global: first they respond to climate change or global warming, which is a global phenomenon. Second, Western environmentalism is local because future of global warming is discussed by legislatures who respond to domestic social problems of concern to a public, which is divided by ideology and interests.

In a sense, the Iraq War has on the one hand weakened the conventional notion of single-issue environmentalism, which is obvious in the case of the land-based Iraqi polity and society and less so in the case of global environmentalism's retreat from single- to multiple-issue movements. Rather than calling No War No Warming a retreat to hybridity, it could also be seen as a sign of waning support for single-issue global environmentalism.

Conclusion

The Marsh Arabs' environmental politics has its own distinctive characteristics; Western concepts of environmental movements, and even environmental politics itself, may not be real equivalents. This distinctiveness partly stems from the militancy of environmental politics, which is not accidental, but the result of the combined effect of authoritarianism, underdevelopment, and marginalization. The conceptual transition from nature to environmental resource through the application of environmental control through engineering is fundamentally a result of the expansion of irrigated agriculture and oil exploitation in the marshes.

In common with other developing countries, violence in the form of armed liberation struggle is a major distinguishing factor between local and global environmentalism due to differences in the political environment within which environmental politics is practiced (Ba'ath authoritarianism versus mature democracies of the West). Indeed, ethnic and minority groups are engaged in broadly defined liberation struggles to protect environmental and natural resources denied to them by the state.

The Iraq war reveals that single-issue global environmentalism is at the crossroads and its survival hinges more on the forms of hybridity it is capable of cementing with other social movements, particularly after the withdrawal of the Coalition of the Willing's forces from Iraq and NATO's imminent withdrawal from Afghanistan in the foreseeable future. The global nature of environmental problems and their linkages with other equally pressing social problems (such as poverty, health, unemployment, education, and other social amenities critical for human survival) clearly challenge global environmentalism in a more fundamental manner than before.

Notes

1. For hybrid environmental movements described by authors elsewhere see Mohamed Salih, *Environmental Politics and Liberation in Contemporary Africa*; Doherty and Doyle, *Beyond Borders: Environmental Movements and Transnational Politics*; and Rojas and Heaney, "Social Movement Mobilization in a Multi-Movement Environment."

2. UNEP, *Environment in Iraq 2001 and 2003,* 5; and Rojas and Heaney, "Social Movement Mobilization in a Multi-Movement Environment," 19.

3. Al-Nasrawi, *Iraq's Burdens: Oil, Sanctions, and Underdevelopment.*

4. UNEP, *Environment in Iraq 2001 and 2003,* Khordagui and Al Aimi, "Environmental Impact of the Gulf War 2007."

5. Iraqi Ministry of Environment, 4 and 8.

6. Iraqi Ministry of Environment, 8.

7. UNEP, *Environment in Iraq: United Nations Environment Programme Progress*; Potocnik, *Environment in Iraq*; Dellapenna, "The Iraqi Campaign against the Marsh Arabs"; Nicholson and Clark, *The Iraqi Marshlands*; and Kelly, "Declaring War on the Environment."

8. "No War No Warming" argues that by controlling Iraqi oil (second-largest oil reserve in the world) and increasing production after decades of sanctions will glut the market with cheap oil, derail government ongoing effort to support alternative green energy, increase fossil fuel consumption, and therefore intensify climate change.

9. Iversen et al., "Gulf War Illness"; Toomey et al., "Neuropsychological functioning of U.S. Gulf War Veterans 10 years after the War."

10. Rojas and Heaney, "Social Movement Mobilization," 12.

11. UNEP, *The Mesopotamian Marshlands,* 14–15.

12. UNEP, *Environment in Iraq,* 2003, and Partow, H. "The Mesopotamian Marshlands: Demise of an Ecosystem," Division of Early Warning and Assessment, United Nations Environment Programme, UNEP Publication UNEP/DEWA/TR.01-3. Nairobi (Kenya).

13. Richardson and Hussain, "Restoring the Garden of Eden," 477.

14. UN Inter-Agency Assessment of Vulnerable Groups, 2003.

15. Maxwell, *People of the Reeds*; Thesiger, *The Marsh Arabs*; Young, *Return to the Marshes,*1977.

16. Maxwell, *People of the Reeds*; Thesiger, *The Marsh Arabs*; Young, *Return to the Marshes.*

17. Thesiger, *The Marsh Arabs*; Cole, "Marsh Arab Rebellion"; and Baram, "Neo-Tribalism in Iraq."

18. Kirstine Adriansen, "What Happened to the Iraqi Marsh Arabs and Their Land?"

19. Cole, "Marsh Arab Rebellion," 6.

20. Baram, "Neo-Tribalism in Iraq," for more on Saddam tribal politics.

21. Coast, "Demography of the Marsh Arabs."

22. Rojas-Burke, "Iraq's Marsh Arabs"; Coast, *The Iraqi Marshlands*; Dellampenna, "The Iraqi Campaign against the Marsh Arabs"; and Kelly, "Declaring War on the Environment."

23. Nicholson and Clark, *The Iraqi Marshlands.*

24. Partow, *The Mesopotamian Marshlands*; and Khordagui, *Environmental Impact of the Gulf War.*

25. Cole, "Marsh Arab Rebellion."

26. Salih, *Environmental Politics and Liberation in Contemporary Africa,* 1999.

27. Clark, *The Impact of Sanctions on Iraq* 1996; Al-Nasrawi, *Iraq's Burdens: Oil, Sanctions, and Underdevelopment.*

28. Richardson et al., "The Restoration Potential of the Mesopotamian Marshes of Iraq."

29. Eden Again Project. Building a Scientific Basis for Restoration of the Mesopotamian Marshlands, 2003, 2.

30. "Interim Report on the Master Plan Released." *Nature Iraq*, Editorial, Vols. 1–2.

31. Iraq Foundation, *The Eden Again Project Mesopotamian Marshland Restoration.*

32. Iraq Foundation, *The Eden Again Project Mesopotamian Marshland Restoration.*

33. The main objectives of the Iraq Foundation are: Fostering the development of institutions of civil society in Iraq through training and education and empowerment programs; working with Iraqi communities and international organizations to develop strategies for Iraq's transition to a pluralist democracy; working with Iraqi civil society organizations to promote human rights education and protections; and promoting the equal participation of Iraqi women and minorities in the public and private sectors of Iraq.

34. Rojas and Heaney, "Social Movement Mobilization," 18.

35. U.S. Green Party, "No to War in Iraq, YES to a Longstanding Peace in the Middle East."

36. Rojas and Heaney, "Social Movement Mobilization," 19.

37. U.S. Green Party, *No War, No Warming.*

38. U.S. Green Party, *No War, No Warming.*

39. UNEP 2001, see also Salmin et al. "Birds: A Key Element of a Biodiversity Survey in the Marshes of Iraq," 2005.

Bibliography

Al-Nasrawi, Abbas. *Iraq's Burdens: Oil, Sanctions, and Underdevelopment.* Westport, CT: Praeger, 2002.

Baram, Azatzia. "Neo-Tribalism in Iraq: Saddam Hussein's Tribal Policies 1991–96." *International Journal of Middle East Studies* 29, no. 1. (February 1997): 1–31.

Clark, Ramsey. *The Impact of Sanctions on Iraq: The Children are Dying.* Bethesda, MD: World View Forum Publications, 1996.

Coast, E. "Demography of the Marsh Arabs." In *The Iraqi Marshlands: A Human and Environmental Study,* edited by Nicholson E. and Clark P, 19–31. London: Politico's, 2002.

Cole, Juan. "Marsh Arab Rebellion: Grievance, Mafias and Militias in Iraq." Fourth Wadie Jwaideh Memorial Lecture. Bloomington, IN: Department of Near Eastern Languages and Cultures, Indiana University, 2008. http://www-personal.umich.edu/~jrcole/iraq/iraqtribes4.pdf downloaded 16 December 2012.

DAI and USAID. Iraq Marshlands Restoration Program Action Plan. Washington, DC: Development Alternatives, Inc. (DAI) U.S. Agency for International Development (USAID), 2004.

Dellapenna, Joseph W. "The Iraqi Campaign against the Marsh Arabs: Ecocide as Genocide." *Jurist: The Legal Education Network,* January 31, 2003. http://jurist.law.pitt.edu/forum/forumnew92.php

Doherty, Brian, and Timothy Doyle. *Beyond Borders: Environmental Movements and Transnational Politics.* Environmental Politics 15, no. 5 (November 2006): 697–712.

Evans, M. I. "The Ecosystem." In *The Iraqi Marshlands: A Human and Environmental Study,* edited by Nicholson E. and Clark P. London: Politico's, 2002.

Iraq Foundation. The Eden Again Project Mesopotamian Marshland Restoration. http://www.iraqfoundation.org/edenagain/publications/pdfs/newslet603.pdf (Accessed December 12, 2011).

Iraq Foundation, Mission, http://www.iraqfoundation.org/aboutus/mission.htm (Accessed January 2012).

Iraqi Ministry of the Environment. Survey on Citizens Views on the Impact of War on Iraq. Baghdad, 2008. In Arabic.

Iversen, A., T. Chalder, and S. Wessely. "Gulf War Illness: Lessons from Medically Unexplained Symptoms." *Clinical Psychological Review* 7 (2007): 842–54.

Kelly, Katherine M. "Declaring War on the Environment: The Failure of International Environmental Treaties during the Persian Gulf War." *American University International Law Review* 7, no. 4 (1992): 921–50.

Khordagui, Hosny, and Dhari Al-Aimi. "Environmental Impact of the Gulf War: An Integrated Preliminary Assessment." In *Environmental Management* 17, no. 4 (2007): 557–62.

Kirstine Adriansen, Hanne. "What Happened to the Iraqi Marsh Arabs and Their Land? The Myth about Garden of Eden and the Noble Savage." Danish Institute for International Studies (DIIS) Working Paper 2004/26.

Maxwell, G. *People of the Reeds.* New York: Harper & Brothers Publishers, 1957.

Nicholson, E., and P. Clark, eds. *The Iraqi Marshlands: A Human and Environmental Study.* London: Politico's, 2002.

Partow, H. *The Mesopotamian Marshlands: Demise of an Ecosystem.* Nairobi (Kenya): Division of Early Warning and Assessment, United Nations Environment Programme. Geneva: UNEP publication UNEP/DEWA/TR.01-3., 2001.

Richardson, Curtis J., and Najah A. Hussain. "Restoring the Garden of Eden: An Ecological Assessment of the Marshes of Iraq." *Geoscience,* 56, no. 6 (2006): 477–89.

Richardson, C. J., P. Reiss, N. A. Hussain, A. J. Alwash, and D. J. Pool. "The Restoration Potential of the Mesopotamian Marshes of Iraq." *Science* 307, no. 5713 (2005): 1307-13–11. http://www.sciencemag.org/content/307/5713/1307.full

Rojas-Burke, Joe. "Iraq's Marsh Arabs." *The Oregonian,* May 14, 2003.

Rojas, Fabio, and Michael T. Heaney. "Social Movement Mobilization in a Multi-Movement Environment: Spillover, Interorganizational Networks, and Hybrid

Identities." February 13, 2008. http://www.hks.harvard.edu/netgov/files/NIPS/Paper_M_Heaney.pdf

Salih, M.A. *Environmental Politics and Liberation in Contemporary Africa.* Dordrecht: Kluwer Academic Publishers, 1999.

Salmin, M.A., A. DouAbul, R. F. Porter, and C. D. Rubec. "Birds: A Key Element of a Biodiversity Survey in the Marshes of Iraq." Paper presented at the 9th annual meeting of the Ecological Society of America, special session 9: Restoration of Mesopotamian Marshes of Iraq; August 9, 2005, Montreal, Canada.

Thesiger, W. *The Marsh Arabs.* London: Penguin, 1964.

Toomey, R., R. Alpern, J. J. Vasterling, D. G. Baker, D. J. Reda, M. J. Lyons, W. G. Henderson, H. K. Kang et al. "Neuropsychological Functioning of U.S. Gulf War Veterans 10 Years after the War." *Journal of the International Neuropsychological Society* 15, no. 5 (September 2009): 717–29.

UNEP. *Environment in Iraq: UNEP Progress Report.* Geneva: United Nations Environment Programme, 2003.

UNEP. "Environment in Iraq: United Nations Environment Programme Progress Report." Geneva: UNEP, 2001.

UNEP. The Mesopotamian Marshlands: Demise of an Ecosystem. http://www.grid.unep.ch/activities/sustainable/tigris/mesopotamia.pdf

UNEP. "United Nations Environment Programme Project to Help Manage and Restore the Iraqi Marshlands." http://marshlands.unep.or.jp/ (Accessed December 2011).

United Nations Inter-Agency. "Assessment of Vulnerable Groups Part I: Marsh Arabs." United Nations Office of the Humanitarian Coordinator for Iraq Lower South, June 2003.

U.S. Green Party. "No to War in Iraq, YES to a Longstanding Peace in the Middle East!" February 2003. http://priceofoil.org/wp-content/uploads/2008/06/warwarming_6x8.pdf (Accessed January 15, 2012).

U.S. Green Party. "No War, No Warming." http://priceofoil.org/wp-content/uploads/2008/06/warwarming_6x8.pd

Young, G. *Return to the Marshes.* Reading: William Collins Sons & Co., 1977.

Chapter 11

Of Causes and Symptoms: Israeli Environmentalism at a Crossroads

Alon Tal

Introduction

The environmental movement constitutes a remarkably successful interest group in Israel and environmentalism is an important part of many Israelis' personal ideology. A 2011 national study confirmed over a hundred Israeli environmental organizations that are active at the local, national and regional levels, with millions of dollars dedicated to staffers, offices and many successful campaigns.[1] The country enjoys a "ruralistic" and Romantic Zionist cultural heritage, which places great value on reclaiming a Jewish connection to its historic homeland.[2] Historically, this is reflected in a strong nature education program in public schools, an impressive program for afforestation and a national passion for hiking and "knowing" the land.[3] National awareness about environmental problems has grown significantly during the country's 64 years of existence. Environmentally oriented parties have registered considerable success at the local, municipal level, and national political polls suggest that there is a potential for Israeli green parties to succeed. Yet, ultimately, an evaluation of the environmental movement needs to rely primarily on "outcomes" rather than "outputs." If Israeli environmentalism is to be judged on the basis of results, the record is less flattering.

Leading Environmental Challenges: While there are certain areas where Israel can claim resourcefulness and improvement, such as forestry[4] and irrigation efficiency,[5] during Israel's 64 years most environmental indicators are distinctly negative. Air quality is largely considered the most daunting public health insult in the country. In 2003, an epidemiological study spearheaded by Israel's ministry of environmental protection along with the U.S. EPA and a local NGO revealed over a thousand premature deaths each year as a result of the fine particles in urban air pollution. For instance, in Tel Aviv some 8 percent of all mortality is associated with exposure to particles while childhood morbidity in Ashdod and Tel Aviv as a result of such exposures is 20 percent and 14 percent respectively.[6]

Despite modest improvements brought about by better industrial performance and the cleaner new vehicles in Israel's fleet, a recent review of air quality by a "blue ribbon" panel of experts, prepared for the ministry of environment, was not sanguine. It concluded that the available data from 2007–2009 reveal a rather grim general picture: Concentrations of air pollutants, such as oxides of nitrogen, ozone and fine particles (particulate matter with a diameter of 2.5 micrometers), are higher in Israel than in Europe and the United States. The levels in the air of Israel's cities of PM 2.5, perhaps the most deadly conventional urban air contaminant, are among the highest in the world, even as about half of the pollutants often arrive as unwanted guests from overseas transboundary natural sources.[7] While some air pollutant emissions appear to be dropping modestly, greenhouse gas emissions per capita are among the higher end of the Western world and they are expected to double by 2030.[8]

Israel's ground water suffers from chronic contamination. The water quality department in Israel's Water Authority reported in 2011 that in Israel's coastal aquifer, the largest single source of fresh water in the country, a full 4.1 billion cubic meters of water—some 10 percent of supply—is contaminated and unfit for drinking.[9] There is a toxic legacy already percolating downward through Israel's lands that even tougher new standards will not be able to address. In 2010, Israel's ministry of environment reported soil contamination around 93 percent of Israel's gas stations, reflected in ground water pollution below 32 percent of the stations. Overall, some 1,195 contaminated land sites have recently been identified across Israel, all constituting potential ground water threats—and this list remains preliminary.[10] Accordingly, the phase-out of many additional wells is a likely prospect.[11]

Only recently have the public health ramifications of exposures to waterborne contaminants been recognized. High levels of pharmaceutical residuals and other "micro-pollutants" are found in Israel's ground water,[12] in some cases as a result of inadequate waste water treatment prior to the massive recycling of effluents.[13] (Israel recycles about 80 percent of its sewage.[14]) These exposures may be part of a broader exposure portfolio involving endocrine disrupting chemicals that are manifested in myriad fertility indicators.[15] For example, a Hadassah Medical Center study showed a 40 percent drop in male sperm counts over the decade preceding 2008.[16] This phenomenon is paralleled by a clearly documented, hormonally related drop in the age of puberty among Israeli girls[17] and a doubling in testicular cancer among Israeli men.[18] Something unhealthy is going on.

Perhaps the most irreversible damage done to Israel's natural resources has been to its open spaces.[19] Israel is among the most crowded countries in the Western world and its physical development has led to considerable sprawl. Already, 5.3 percent of the country's 22,000 square kilometers is "built." But the percentage is far higher when the unpopulated desert regions are left out of the calculation. For instance, in the greater Tel Aviv planning region, over 65 percent of lands are developed; 11 percent of the lands in the northern half of the country are covered; 130 square kilometers are paved over by roads.[20] Planners have been unable to stem the tide—between 1998 and 2007, 83.5 square kilometers of built space was added with a parallel subtraction in open spaces.[21]

This anthropogenic incursion into local habitats has caused massive damage to Israel's rich natural world. Global surveys have identified Israel as an international "biodiversity hotspot."[22] For instance, Israel offers a home to 115 wild mammal species, more than half the number existing in California, which is 14 times its size. But of these, 25 of Israel's mammal species are endangered and 12 more are defined as *critically* endangered. A mere 30 mammal species enjoy stability and are not defined as threatened at all.[23] After years of aggressive water management, only 3 percent of Israel's remaining wetlands remain intact, and the country's amphibian population literally has no place to go. Of the six indigenous species, all are endangered as are a third of Israel's 103 reptile species.[24]

Similar failures could be highlighted in the areas of solid waste, energy conservation, pesticide usage, and exposures associated with the highest cell phone usage in the world. Many iconic local sites are simply disappearing, from the already drained Huleh Lake to the vanishing Dead Sea to Israel's once-rich and colorful coral reefs in the Gulf of Aqaba.

Chapter outline: These figures offer a grim context for assessing the impact of environmentalism and Israel's environmental movement's performance and influence on political decisions to protect the country's natural resources. While it is possible to offer a highly self-congratulatory view of the remarkable achievements of Israel's civil society and the response of individual politicians, this would seem to be missing the discouraging "big picture."

The question that this review will consider therefore is this: *How is it that a robust and ostensibly effective environmental movement has not succeeded in producing meaningful environmental progress? What changes need to happen for Israeli environmentalism to become more effective?* After briefly describing the dynamics of Israel's political and legal systems in the context of responsiveness to environmental protection, three reasons are presented for Israel's failure thus far in galvanizing political power and public involvement to reverse the pervasively negative environmental trends:

1. Israel's rapid population growth has undermined conservation and environmental protection efforts; neither Israel's politicians nor its environmental community have been willing to address these dynamics.
2. Israel has emerged as a consumer society that has created consumption trends that exacerbate the ongoing environmental degradation.
3. Israel's environmental problems remain an issue that is perceived as offering a low political return, making interventions by politicians idiosyncratic and preventing the national presence of a powerful green party.

Political Dynamics of Israeli Environmentalism

Introductory Civics: Israel is a parliamentary democracy. Although the country has no formal constitution, a series of basic laws provide a democratic framework for a representative, "elected" government. The country's Declaration of Independence from 1948 offers a strong vision of a country that protects human rights and aspires to a pluralistic vision of equity, where Jewish and democratic values happily coexist.[25] It is not surprising that this "level playing field" offered a hospitable habitat for the steady growth of an environmental movement.

Israel's parliamentary system relies on open elections every four years to all citizens where 120 representatives are elected to Israel's Knesset. Based on the results, Israel's president grants the party with the greatest chance of forming a parliamentary majority the opportunity to form a government.[26] Parties are required to attain 2 percent of the votes before they are awarded seats in the parliament, a relatively modest threshold by international standards. Although there has been an increase in the role of "personalities" of prime ministerial candidates,[27] most people still vote according to ideological lines or traditional patterns and ethnic community loyalties. Accordingly, there are many parties in Israel representing a rich diversity of the country's diverse demographic mosaic.[28] The upshot of the system is that the government increasingly relies on the support of small parties to ensure its parliamentary majority. This frequently gives small parties disproportionate influence on government policies.

Green Politics: In theory, the environment is a value shared by all segments of Israel's disparate population. And indeed, there have been a broad range of politicians who have shown interest in promoting legislation to protect the environment, lobbying the government and the parliament to ensure that public policy is "pro-environmental."[29] But while green parties have run for election in Israel's Knesset four times, they have not cracked the requisite threshold, even as they came very close to doing so in 2006.[30] They invariably garner more than 1 percent support of the votes, but this is not enough to break into the parliament.[31] Presumably, existential issues such as security or religious issues continue to trump lower priority matters such as environmental health and nature preservation.

In contrast, municipal elections in Israel have provided a successful platform for environmental parties. Today, some 40 representatives of environmental parties serve in Israel's city councils.[32] In the major cities of Tel Aviv, Haifa, and Jerusalem, as well as smaller towns such as Modi'in, green factions were sufficiently large to negotiate "deputy mayor" positions from the ruling mayor as part of coalition agreements. In the context of local governance, where Palestinian borders and Iranian nuclear capabilities are not salient issues, there is a growing number of Israelis who see the environment as a compelling issue worthy of their attention. While municipal elections offer the best empirical proof of the willingness of tens of thousands of Israelis to prioritize environmental issues in their electoral decisions, it is still a relatively small minority that put these challenges at the top of their personal, political considerations.

Civil Society: Israel's civil society probably offers the most widespread expression of the public's "love of the land" being manifested as a strong, collective impulse among Israelis. A 2011 study identified over 100 active environmental organizations in Israel. Several have annual budgets exceeding a million dollars and it is likely that all told, some 100,000 have formal affiliation with some environmental agency or group. The largest environmental NGO, the Society for Protection of Nature in Israel (SPNI), boasts over 40,000 households among its members. For a country this size, environmentalism is hardly a trivial force to be reckoned with.

Tactically, Israel's environmental movement has always managed to balance the tactical dilemma of whether to play an insider role or to be an outside force for change. In the 1950s, during its formative years, the SPNI assumed a clear insider role, receiving considerable government funding and focusing efforts on friendly lobbying of its many colleagues holding senior governmental positions rather than assuming a confrontational approach. Yet, ever since an influential mass gathering in 1962 to oppose the development of the Carmel Mountains was dubbed by the local press as a "demonstration," the organization has enjoyed playing both roles. Like other NGOs, there are times when it chooses to sit as a member or an observer on government committees. In other cases it sues the government in the Supreme Court. Often organizations will coordinate, assigning different roles—with one cooperating with decision makers as a "good cop" and the other protesting in the streets.

As expected in any complex "niche ecology," the myriad organizations working on green matters are diverse in terms of size, objectives, approaches, and activities. And yet, they work surprisingly well together in coalitions, with dozens of ad hoc and several formal alliances established to conduct a particular campaign or to coordinate work around a given issue. This remarkable unity can be attributed to the many values the environmental NGOs hold in common and a largely shared ideological orientation.[33]

Environmental leader Eilon Schwartz has charted an evolution in the ideology of Israel's environmental community in an analysis that enjoys considerable recognition among green activists.[34] Judaism has much to say about stewardship and humility in human relations to creation.[35] While cognizant of this tradition, the founders of Israeli environmentalism were not religious people and were largely driven by other, universal intellectual trends. Schwartz explains that initially, from the 1950s through the 1970s, there was a focus on nature preservation. In these days, the Society for Protection of Nature, Israel's oldest and largest environmental NGO, led innumerable campaigns to protect venerated landscapes and local biodiversity. A Nature Reserve Authority, established in 1963, led government efforts to declare a network of nature reserves that now includes 25 percent of the country's lands.

Eventually, as Israel became a more urbanized society and egregious hazards and nuisances began to interfere with the quality of day-to-day living, there was mounting concern for public health. At the civil society level, originally during the 1970s, such work was only undertaken by MALRAZ—an acronym for the National Council for Prevention of Noise and Air Pollution. By and by, during the 1990s, a broader community of local grassroots organizations emerged, along with the public interest law and science advocacy group, Adam Teva V'din, the Israel Union for Environmental Defense, that brought a greater professional presence to a full range of environmental media and problems. The establishment of the ministry of environmental protection in 1988 was a critical step in establishing a national regulatory program to promote these ends.

Evolving Environmental Ethics: During the last decade or two, environmentalists began to embrace a more integrated approach to environmental issues, which focuses on the broader social implications of Israel's environmental crisis. Schwartz calls this stage "place-based environmentalism." Indeed, during this period, concerns for environmental justice[36] and the gap in access to resources and quality of life with Israel's poorer and minority communities has emerged as a central concern among Israel's environmental organizations. Many consider the tremendous public awareness generated in an unsuccessful campaign to prevent the paving of Highway 6 as a turning point in Israel's environmental movement's awareness and willingness to take on a broad variety of issues and partners.[37] Access to sustainable public transportation is an issue that not only involves air quality and open spaces, but also concerns for social equity and public property, offering a classic opportunity to bring social and environmental values together. Table 11.1 offers a review of the components of these three stages.

As their values and orientations evolved, environmental public interest groups have consistently been the engine behind environmental progress in Israel. Initially, they began by lobbying senior decision makers in the political establishment, often personal cronies and contemporaries who then dominated the Ashkenazi (of European origin) establishment in the ruling leftist political parties. Later, during the 1970s, they became more confrontational, initiating a series of high-profile demonstrations[38] with a greater emphasis on legal intervention in more recent years.[39] Indeed, the highly permissive attitude of Israel's courts to public petitioners and a series of statutory amendments granting standing to environmental groups have created an impressive legacy of public interest litigation legal precedents where NGO petitioners facilitated judicial intervention to protect environmental values.[40] Environmental education slowly became part of the national curriculum in public schools.[41]

Israel constitutes an extreme example of a multicultural society.[42] It can surely be argued that along with the expansion in the environmental movement's ideological orientation came a broader support base across the splintered Israeli public. Although, initially, Israel's environmental movement was largely popular among Israel's solidly middle class, secular Ashkenazi establishment, recently there has been greater interest among a

Table 11.1. Elements of the Three Stages in Modern Israeli Environmentalism

Paradigm Characteristics	Nature Conservation	Environmental Sciences	Place-Based Environmentalism
Terminology	Nature conservation	Environmental quality	Lifeways, environmental practices
Focus of Action	Nature reserves	Public health, safety	Social-environmental quality
Environmental Problems	Biodiversity, open spaces, indigenous peoples	Air and water pollution, dwindling resources	Urban life, alienation, loss of quality of life
Human-Nature Relationship	Nature as a spiritual source, apart from humans	Humans and nature belong to the same system (lack spiritual dimensions)	Humans and nature mutually interconnected (physically and spiritually)
Moral Approach	Biocentric	Anthropocentric	Cultural anthropocentrism (human culture rooted in nature)
Academic Fields	Ecology, biology	Environmental studies (soil, water, policy)	architecture, philosophy, history, agriculture
Basic Concepts	Deep ecology	Sustainable development	Bio-regionalism
Educational Approach	Sensory, emotive, experiential (wonder, nature walks)	Cognitive-analytical; environmental literacy, laboratory, research policy	Relation to place local knowledge, identity, cultural studies, empowerment

Source: Schwartz, 2002, Three Paradigms of Environmental Education in "Paths to Sustainability," p. 38.

variety of ethnic groups including Sephardic communities (Jews from Arab countries)[43] as well as involvement among Arab Israeli citizens and other religious groups.[44]

Government Regulation and the Environment: The efforts of the ministry of environmental protection and other government agencies should not be dismissed or minimized when evaluating Israeli environmental progress in the last 20 years. Yet, the actual influence of the ministry in key controversies was frequently dependent on the particular talents and commitment of the minister. With only a modest budget, the minister's post has always been considered a low-prestige appointment. Indeed, when Labor politician Daliah Itzik received word that she was to serve in the position in Prime Minister Ehud Barak's cabinet she openly pouted to the press that she had been punished and given a "booby prize."[45]

There have been particularly impressive ministerial tenures by politicians, such as Leftist Meretz party leader Yossi Sarid during the 1990s or the present tenure of right-wing Likud politician Gilead Ardan. Yet, ultimately these did *not* reflect a trend that upgraded the ministry of environment, transforming it into a more influential and effective player. Without the power of a political party committed to many of the ministry's objectives, the environment ministry with its limited authorities and staff found that it often lost key battles to other, more formidable ministries' agendas. Hence, in critical areas such as transportation policy, planning and open space protection, energy conservation, mining, sewage treatment, drinking water regulation, clean electricity and so forth, the ministry at best can be considered a supporting actor rather than a leading protagonist in formulating public policies.

Israeli Environmentalism—Summary: The previous two examples of dedicated politicians who headed the environmental ministry are ultimately anomalous. But they do reveal an interesting dynamic characterizing the politics of Israeli environmentalism. Unlike the United States, where the two major parties have recently come to represent very different approaches to environmentalism,[46] in Israel it is difficult to characterize the "Right" or the "Left" as particularly pro-environmental. In a review of Israeli parliamentarians who have crusaded for environmental causes over the past 40 years, Open University expert Orr Karassin concludes:

> These true believers have not come distinctively from parties associated with the left, and have forged unique political alliances and cooperation on environmental issues. They have demonstrated environment to be a non-partisan agenda capable of producing one of the only consensual goals in Israeli politics.[47]

In summary, Israel's environmental movement has all the elements that should lead to sensational success:

- A strong cultural tradition that values harmony between the residents and the land
- A growing public awareness of Israel's environmental problems
- A dynamic environmental movement that has grown and whose perspective has grown over the years
- A legal system that is uniquely open to public interest, environmental litigation
- A political system that is open to environmental advocates and where for decades the environment has remained a nonpartisan issue

Add to this the fact that Israel's economy has steadily grown stronger over the years. Today the country enjoys the collective financial resources to make a serious budgetary commitment to environmental issues and regulation. And yet, the environmental indicators described in this chapter's opening section are largely negative. So it is well worth considering in greater depth the reasons behind this disappointing performance.

Analyzing the Strategic Shortcomings of Israeli Environmentalism

Israel's environmental movement has produced many local successes. It has written and passed environmental laws; it has saved countless natural treasures around the countryside; it has helped educate a nation regarding the seriousness of present environmental challenges. But significant, systemic reversal of negative trends continues to elude it, leaving Israeli environmentalism's record a mixed one. There are many reasons that can explain the present dynamics, but three are particularly important:

1. The relentless growth in Israel's population
2. The transition to a consumer society
3. The lack of significant political muscle promoting an environmental agenda

The politics surrounding these three factors will be described in greater detail.

Overpopulation

During the turbulence of the War of Independence, most of the Arab majority left the country and massive Jewish immigration ensued.[48] Accordingly, by 1950 there were a million people living in Israel.[49] After six decades with an average annual demographic increase of 3.6 percent, today there are 7.8 million permanent residents.[50] The steady growth in Israel's population is presented in Figure 11.1.

Figure 11.1 Population in Israel, 1950–2020

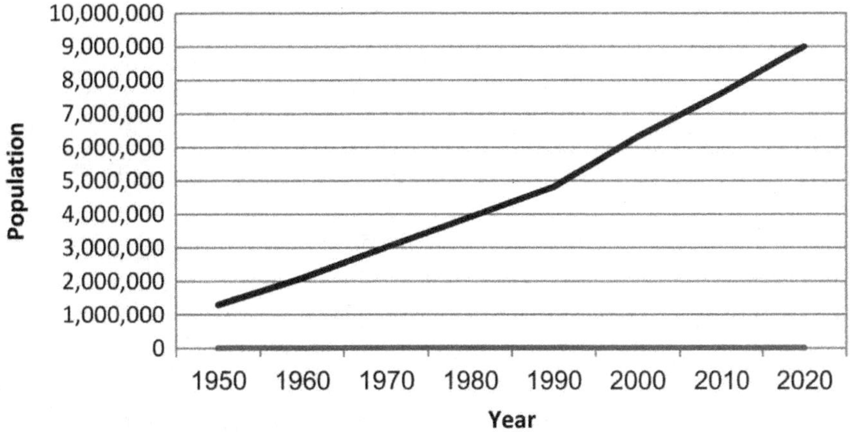

Source: Israel Statistical Bureau, 2007.

Growth Rate: To understand how this growth is actually magnified in terms of effective population density requires an understanding of Israel's asymmetrical geography. The south of Israel, generically called the Negev Desert, covers 60 percent of the country's landmass. But only 8 percent of Israelis live there. Population density for the entire country is roughly 320 people per square kilometer, 10 times the average in OECD countries.[51] But in practice, 92 percent of Israelis live with a population density of 710 people per square kilometer.[52] Present trends will push that number to 900 by 2020. By way of comparison, even a very crowded country like the Netherlands has only 400 people per square kilometer; Belgium and Japan have 354 people per square kilometer and 337 people per square kilometer respectively. As anyone stuck in a traffic jam can attest, the effect of crowding on quality of life can be adverse.

Environmental Implications: Of course the environmental implications are also enormous. Israel's water shortage is increasingly acute, with the Kinerret Lake, Israel's only meaningful surface water reservoir, at its lowest point in history.[53] Chronic water scarcity is exacerbated by the climate change that has caused a statistically significant drop in precipitation rates.[54] But Israel, which might be a world leader in solar energy, has not made a commitment in this area: the country's greenhouse gas emissions continue to increase due to the dramatic rise in electricity demand that averages 4.5 percent a year.[55] Solid waste generation in Israel also steadily rises while the available sites for disposal are filling up.[56] The physical development required to meet the exponential population growth has been described earlier, but it is the primary driver behind the deterioration in open spaces and the disappearance of ecological corridors across the country.[57] Population growth also exacerbates environmental disputes. When Hebrew University geography professor Eran Feitelson undertook a review of the major environmental conflicts, he discovered that the majority of them reflected situations where development expanded toward existing neighborhoods, with significant environmental health ramifications and tensions.[58]

Israeli environmentalists are increasingly keen to enumerate the social implications of environmental problems, and Israel's rapid population growth offers an extreme case in point. The present affordable housing crisis, the shortage of hospital beds, the overcrowded classrooms, the considerable delays in getting court dates, the ubiquitous traffic gridlock, and the growing vulnerability in food security—all can be attributed to the astonishingly high rate of population growth. But no single area reflects the scourge of present demographic dynamics more than poverty. The percentage of Israelis living below the poverty line has grown from 25 percent in 1979 to 33 percent in 2010—an alarmingly high number for a country with a commitment to social equity.[59] The vast majority of poor families are large families. Educational achievements among children in Israel's large families are far lower than those living in families with two to four children.[60] In summary, like the proverbial 10 blind men, who all try to describe an elephant and after touching different parts, depict completely different phenomena—these social afflictions are all features of the same problem: the elephant in the room is population growth.

Causes of Population Growth: There are two engines for the growth in the number of Israeli citizens: one is the immigration of Jews from around the world and the other is the high birth rate. Jewish immigration is facilitated by the "Law of Return," one of the country's first statutes, which welcomes Jews from around the world to come and claim their birthright as Israeli citizens. The law is one of Zionism's axiomatic statements, reassuring Jews around the world that after millennia of persecution, they will always have a country where they can find refuge.[61] Yet, with the apparent end of mass immigration from the former Soviet Union, immigration levels have dropped. Today, the growth in population is primarily driven by the large average size of Israeli families.[62]

While clearly there is a cultural/religious component to these dynamics, a closer look suggests that the high population growth rate is the result of public policies and "perverse incentives" that both encourage large families and signal the heads of the households that they do not need to work. Child allowance subsidies in Israel are designed to increase with each additional child. This sends a clear message: "more is better." The overall increase in these subsidies and the upwardly staggered rates were adopted as a concession to religious parties. As a result, since the 1970s, while Israeli GNP increased 100 percent, subsidies for children went up by 540 percent.[63]

The direct payments by the National Insurance Institute to large families is in fact just the tip of the iceberg. Other subsidies can be found in policies that offer special government cash bonuses that women receive after giving birth, significant reduction in expectations for military service and reserve duty for fathers of six children, as well as exemptions for pregnant women and exemptions from city property taxes for large families. Even fertility treatments, which cost some 12,000 shekels per case, are offered as a free health service, regardless of a family's size.[64] In short, hundreds of thousands of Israeli households enjoy thousands of shekels of remuneration each month, which allow them to maintain a large family. Before these massive pro-natal subsidies, ultraorthodox families—which today have an average of six children per family—had birth rates comparable to those of the general public.[65] And when in 2003 the ministry of finance made a modest reduction in the level of childhood subsidies, there was an immediate drop in the birth rate. The drop was greater in Israel's Arab sector than among ultraorthodox women.[66]

Present Policy Discourse: The legitimacy of these public policies is not a part of the public discourse in the environmental world, but it should be. In the 2011 survey, some 55 percent of environmental organizations agreed that it was time that Israel's environmental movement confronted the country's demographic/population dynamics. In fact, less than 3 percent of environmental organizations report that the issue is actually on their organizational agenda and it is difficult to identify a single significant campaign—political, educational, or otherwise—among local NGOs that highlights Israel's unsustainable demographic trends.[67] Israeli environmentalism is not yet "on board."

The reasons for the reticence are clear. Environmentalists are concerned about their image and fear that they might be caricaturized as fanatical or out of touch with the national consensus. Raising the question of family size offends the politically correct etiquette of enlightened society and is often censured as an expression of anti-Arab or anti-ultraorthodox prejudice. As so many of the children in large families already "go to bed hungry," a call for policy reform is attacked as obtuse at best, and intentionally cruel at worst. Some organizations are sensitive to what they believe are the proclivities of their funders, for whom overpopulation is not an environmental priority. And there are those who explain that until the Jewish people replace the 6 million victims who died during the World War II Holocaust, there is no place for such discussions. Indeed, in an internal debate, a 2011 convention held by Israel's green party, The Green Movement, decided to recognize the significance of the population as an environmental and social issue but to downplay its immediacy as an election issue in order not to jeopardize public support.

Of course a comprehensive policy response to the population issue requires solutions that go far beyond the comfort zone of conventional environmentalism. They include empowering women in the workplace, providing access to education and birth control, reducing obstacles to abortions as well as providing real incentives to families that contribute to a stable population level.[68] Such policies require a change in the ethos of Zionism, which for many years saw increasing population as a paramount national defense and economic objective. Religious communities in Israel, which are inspired by the Biblical call to "be fruitful and multiply,"[69] are particularly resistant to reconsidering the environmental and social implications of present demographic trends. But in a world with finite resources, "sustainable growth" is ultimately an oxymoron.

If Israel's population should double yet again, it is impossible to imagine real progress in any of the country's disturbed environmental media. Environmental organizations have been willing to voice inconvenient truths in the past. If Israel is ever to reach an equilibrium that promises quality of life for its citizens and the creatures of the land, it is time they embrace the issue and introduce a new vision of zero population growth.

Consumption and the Environment

Israel for many years was considered a developing country. Indeed during its early years, basic commodities like eggs were rationed and per capita GNP was well below $3,000. But that situation changed as the country steadily became more prosperous. After several decades of economic growth that averaged between 3 percent and 4 percent, Israel is today a wealthy nation and in 2010 acceded to membership in the OECD. With a per capita GDP that has reached $30,000 a year, the country's economy was surprisingly unaffected by the recent global economic turmoil.[70] That of course is a good thing. But it's not clear whether the new wealth being generated is being equitably distributed or is always being used for the right things.

Israeli society is being swept by a wave of consumerism. Children spend an increasingly significant part of their free time in malls and shopping for many has become an obsessive form of recreation. As a country that used to pride itself on frugality and humility, conspicuous consumption is considered by many to be a status symbol. The resulting production and waste have profound environmental implications.

Environmental Impacts of High Consumption: Solid waste is a classic example. Israel's ministry of environmental protection reports that the amount of garbage generated nationally increases at an average rate of 5 percent a year, doubling quickly during the past two decades.[71] Today per capita solid waste consumption is rapidly approaching 700 kg per person per year—still behind the United States level of 730, but way ahead of the 430 European levels.[72] Even as the country seems to have put policies in place to expand recycling activities, such as landfill taxes,[73] the sheer volume of rubbish disposed off every day will remain a problem for the limited carrying capacity of a small country.

Another area where consumption impacts environmental quality is the increasing level of motorization. There are now about 2.5 million motor vehicles in Israel, about 80 percent of which are private. In terms of trends: in 1990 there were 211 vehicles per 1000 people, and this level rose to 288 in 2000 and up to 326 today.[74] This is not just a reflection of dispersed planning paradigms, which force the public to buy cars in order to maintain minimal levels of mobility. Among certain circles in Israel, cars have also become a status symbol, coming at the expense of the veneration of public transport alternatives. Housing size is also growing. A recent study showed that the average size of new apartments grew from 140 square meters to 200 square meters.[75] While many Israeli apartments are cramped, especially for large families, the increased expectations for spacious living quarters ultimately will affect the available open spaces, as suburban neighborhoods continue to exacerbate sprawl patterns around the country.

Of Values and Political Will: The environmental crisis is ultimately a crisis of values. In a free society, beyond the taxing of luxury goods, it is probably both inappropriate and politically impossible to constrain consumption. Yet, a majority of Israelis sense that the country needs to send a clear message about the vulgarity of conspicuous consumption and that meaning in life will not be found in compulsive acquisition of "things." In the past, the Israeli public has proven itself extremely open to environmental messages about the public good. For example, after three consecutive drought years, a severe water crisis hit Israel in the summer of 2009. The Israel Water Authority launched a massive media campaign calling for individual conservation. The results were remarkable, with an immediate 13 percent voluntary drop in consumption levels by the general public.[76] The same message could be brought to issues of consumption in general, especially in a country that prides itself on moral integrity and "doing the right thing."

What's lacking is political will, civic discipline, and societal commitment to consider consumption as a public policy issue. Rather, the environmental movement focuses on the end of the pipe solution—for example, recycling—but does not ask the more difficult questions about why there are so many things to throw away in the first place. Green Course, the national student environmental network, intermittently promotes the international "Buy Nothing Day." But one can argue that the Jewish tradition, with its abstinence from commercial activities on the Sabbath, already has a "buy nothing day" every week. In an attempt to inculcate green values among the public and integrate Jewish tradition into modern life, the Green Movement political party has endorsed a Sabbath policy that would continue to keep commercial establishments closed on Saturday, while leaving cultural venues, restaurants and the like open to serve the public. Such a position is considered sanctimonious by many Israeli environmentalists who are not keen on preaching a gospel of greater simplicity.

Raising the issue of consumption and its effect on the environment is a hard sell anywhere.[77] People do not like to hear that they are contributing to a serious problem, and rarely are they happy to change their lifestyles. But this does not diminish the truth about the planet's finite resources. It seems that too often, environmental advocates everywhere choose a path of least resistance. While heavy on "lip service," in practice, environmentalism in Israel does not yet have consumption on its operational agenda.

Environmental Issues as Political Orphans

Over the years, environmental organizations have been quick to avail themselves of the Israeli political process in promoting their agendas. Indeed, most of the key environmental legislation adopted by Israel's parliament during the past 20 years was prepared by NGOs and conveyed to legislators who submitted them as private legislation. An interesting related phenomenon is the high number of environmental laws that are passed literally hours before the Knesset recesses before elections. In the anarchy that accompanies the dissolution of the government, it is possible to circumvent the usual constraints imposed by the coalitional discipline and appeal to parliamentarians' sense of higher purpose. In this way statutes involving coastal protection, public participation, marine pollution, and even the most recent Clean Air Law, all managed to "slip through" the legislative process, and under the radar, receiving the requisite votes in parliament. Of course this would not be possible without arduous work done in the corridors of power by the so-called "green lobby."

State of Israel's Environmental Movement: The recent environmental survey reported that over 10 percent of Israeli NGOs invest substantial resources in lobbying activities—with the major organizations hiring full-time lobbyists or outsourcing lobbying activities to professional lobbying agencies.[78] Key to recent legislative successes is the internal lobby of 18 elected parliamentarians from across the political spectrum who have committed to working on behalf of environmental causes.[79] In fact, the parliament's present group is relatively small compared to past sessions of the Knesset.

The lobby serves as a catalyst for nonpartisan initiatives. For example, recently 70 of the 120 members of the Knesset signed on to a bill that would provide support for bicycle riding in Israel.[80] Working through their umbrella organization, Life and Environment, prizes are awarded in a major Tel Aviv ceremony by NGOs each year to outstanding "green parliamentarians" along with a "black globe" award for anti-environmental villains.[81] In an era when many parties require its members to compete in primaries to attain a seat in the Knesset, these events matter. The implications of such highly publicized partnerships are not insignificant; many politicians may hesitate before antagonizing the environmental community.

Nonetheless, in numerous cases, the environmental movement has lost major campaigns because it could not wield the political muscle to overcome opposing interest groups. Such weakness was readily apparent in recent failed efforts to stop the privatization of public lands, prevent the dismantling of the physical planning system in a new law that allows for quick (and presumably environment-unfriendly) decisions, or cancel the quarrying of the Samar sand dunes. Indeed, Israel's most contentious environmental controversy to date, involving the paving of Highway 6, ended in a total rout, and the road was built expeditiously.[82] In most of these losing efforts, the minister of the environment was an avid partner with the environmental NGO community, but could not convince his colleagues in the cabinet to take a pro-environmental stand.

While all parties like to paint themselves as environmentally enthusiastic, the truth is that few have a real interest in taking on environmental causes that involve confrontation; most simply do not see the environment as a compelling political issue. Their polls tell them that at the end of the day, taking a position that damages the environment will not affect them at the ballot box.

Government coalitions in Israel are comprised of small parties. Because they hold the balance of power in the tenuous power struggles that invariably result from Israel's inconclusive elections, many of them succeed in attaining significant concessions to promote their narrow interests. These narrow objectives can range from increased funding for religious institutions to supporting the demands of farmers or retirees. Sadly, the environment, a public good for all, has never enjoyed this sort of patronage from Israel's political parties. Individual leaders may make a good case in cabinet debates, but never has the coalition's stability been threatened over the outcome of an environmental issue. Nor has any party ever made environmental commitments an item in the negotiations leading up to the formation of a government coalition.

Such dynamics are often different at the municipal level. This is because more and more Israelis with environmental inclinations have entered the fray of city council politics. Frequently, green party representatives succeed in joining the ruling coalition and manage to coax concessions and even budgetary commitments from the mayor. Not all demands are met, but in a democratic system, when a city council member has made her environmental positions the basis for her election, basic notions of accountability motivate her to see that there are tangible achievements during a five-year term in local government.

It would seem that in order for such a quantum leap forward for environmental agendas to exist at the national level, a change in the political mosaic of the country will be necessary. This implicitly means that a green party must either become a meaningful faction in a major existing party or run independently and integrate itself into a ruling coalition. Experience in Europe—particularly in countries like Germany and Finland where green parties joined the cabinet—have shown the extraordinary achievements that are possible under such conditions.[83] Whether there are enough Israelis who view the country's environmental and related social issues as sufficiently important to justify voting for an Israeli green party in national elections remains an open question.

Conclusion

In their classic evaluation of modern environmental problems, Ehrlich and Holden posit an *impact law.* Typically depicted as a simple equation,

E = P+A+T, it holds that *Environmental damage* is a function of *Population, Affluence* (consumption), and *Technology.*[84] An assessment of the political dynamics of environmentalism in Israel suggests that while there has been an impressive series of isolated successes, they have not resulted in sustainable trends. The local environmental movement and the politicians who are associated with environmental causes typically have perspectives and agendas that are limited to the third element of the Impact equation—*Technology. Population* and *Consumption* simply are not part of their conceptual universe.

As a result, Israel's environmental movement finds itself addressing symptoms rather than causes. Oftentimes, the heroic efforts of Israel's public interest NGOs are likened to the proverbial Dutch boy, who stuck his finger in a hole in a valiant effort to stop a breach in the dike. With limited fingers, it is not clear whether the environmental community in Israel can field an effective response to every crisis. Advocates speculate that perhaps the existing "dike" needs to be replaced with something more formidable. But relating to the metaphor in this way is in fact backward—as a dike constitutes a response to a vexing, external threat. Replacing it will not get to the heart of the problem. Israel's growing population and insatiable consumer culture are more analogous to the sea, whose level is rising and will continue to grow fiercer with time until it overwhelms the carrying capacity of the country. A real environmental solution must confront these problems head on. Environmentalists are not asking the right questions, no Israeli political party has a fundamental interest in engaging the Israeli public in a real discussion about the direction in which the country is heading and its dwindling natural and environmental resources.

This is not to say that these dynamics will last forever. Israeli environmentalism is at a crossroads. With a strong heritage that has always actively sought to redeem its ancient homeland,[85] broad public support and impressive institutional capacity, it now needs to expand its operational orientation and objectives. The fact that a majority of environmental groups today acknowledge the importance of the population and consumption issues represents a profound change and break from past taboos.[86] Academics can also make some noise and do the necessary research to build the case linking environmental degradation to overconsumption and overpopulation. But, it is not clear that such efforts alone will succeed in challenging core values of Israeli society. Much like the situation that has transpired across Europe, a political intervention will be necessary for environmentalism to catalyze meaningful progress.

Sadly, security will likely remain the most compelling issue for Israeli voters in the foreseeable future. Yet, it is also possible that a growing number of citizens will come to recognize that none of the sundry governments that have ruled the country over the years have been able to appreciably improve the prospects for a secure peace. One inference from such a conclusion is that votes might be better used if they focus on domestic issues. Here, presumably, there is a greater ability to influence public policy.

Israel's first prime minister, David Ben-Gurion, was fond of saying that in Israel the very difficult is done tomorrow while the impossible will take a little longer. The present environmental challenges of Israel are daunting. Presumably Israeli environmentalism is strong, and the political and legal infrastructure exists for implementing comprehensive solutions. But in order to succeed, the real causes of the country's environmental crisis need to be addressed and advanced by a formidable political force. This may take a little longer.

Notes

1. Tal, Leon-Zchut, Frankel-Oshri, Greenspan, and Akov, *Israel's Environmental Movement.*
2. de-Shalit, "From the Political to the Objective," 81.
3. Tal, *Pollution in a Promised Land.*
4. Ginsberg, "Afforestation in Israel, A Source of Social Goods and Services," 43.
5. Tal, Ben-Gal, Lawhon, Rassass, *Sustainable Water Management in the Drylands.*
6. Preuss, Goren, Haran, Nezer, and Warburg, "A Comparative Assessment of Air Pollution Public Health Risks in Two Israeli Metropolitan Areas—1995 to 1999."
7. Public Expert Committee Regarding the State of Air Quality in Israel, *National Masterplan, Clean Air Law,* 4–5.
8. McKinsey and Company, *Greenhouse Gas Abatement Potential in Israel.*
9. Rinat, "Green Groups Warn."
10. Israel Ministry of Environmental Protection, "Environmental Protection Ministry Publishes Indicators on the State of the Environment in Israel."
11. Rinat, "Contamination Endangers Coastal Aquifer."
12. Lev, "Regulated and Non-regulated Microcontaminants in Drinking Water and Wastewater in Israel."
13. Avisar, Lester, and Ronen, "Sulfamethoxazole Contamination of a Deep Phreatic Aquifer," 4278–82.
14. Tal, "Seeking Sustainability" 1081–84.
15. Zafrir, "International Expert."
16. Har-Nir, et al. "Declining Quality Of Donated Semen."
17. Flash, "Long-term Secular Trends in the age at Menarche in Israel."
18. Matzkin, "Testis Cancer in Israel."

19. Tal, "Space Matters," 119–51.

20. Kaplan et al., "Patterns in the Utilization of Constructed Land in Israel."

21. Israel Environmental Protection Ministry, "Ministry Publishes Indicators on the State of the Environment in Israel."

22. Myers et al., "Biodiversity Hotspots for Conservation Priorities," 853–58.

23. Shalmon, "Mammals in Israel."

24. Bouskila, "Reptiles in Israel."

25. See former Israeli chief justice Aharon Barak's review of this issue in: Barak, "The Values of the State of Israel as a Jewish and Democratic State."

26. For a review of Israel's constitutional government see: Susie Navot, *The Constitutional Law of Israel.*

27. Rahat and Sheafer, "The Personalization(s) of Politics," 65–80.

28. Galnoor, "Crisis in the Israeli Political System," 144–75.

29. Karassin, "The Battle of the 'True Believers'," 168–89.

30. Election results in English can be found in "Israeli legislative election, 2006."

31. See website of Israel's green party "The Green Movement" at www.yeruka.org.il

32. See description of the "Green Municipal Forum" in the Green Movement NGO website: www.green-ngo.org

33. Tal, *Israel's Environmental Movement.*

34. Schwartz, "Appendix I: Three Paradigms of Environmental Education," 38.

35. Benstein, *The Way into Judaism and the Environment,* 51–53. See also: Waskow (ed.), *Torah of the Earth, Exploring 4,000 Years of Ecology in Jewish Thought.*

36. Lubinov, *Environmental, Social and Economic Justice and Human Rights.*

37. Meizlish, *The Battle against the Trans-Israel Highway*; also Rabinowitz and Vardi, *Driving Forces.*

38. de-Shalit and Moti, "Green or Blue and White? Environmental Controversies in Israel," 273–95.

39. Noga Morag-Levine, "The Politics of Imported Rights," 334–53.

40. Marom-Albeck and Tal, "Towards a More Effective Model of Public Interest Environmental Citizen Suits," 373–424.

41. Negev, Sagui, Garb, Salzberg, and Tal, "Evaluating the Environmental Literacy of Israeli Elementary and High School Students," 3–20.

42. Tamir, "Shnei Musagim Shel Rav Tarbutiyut (Two Concepts of Multiculturalism)," 79–92; also Shenhav, *What is Multiculturalism?*

43. Sosner, *The Environmental Campaign as a Force for Molding a New Identity.*

44. Negev, Garb, and Benstein, *Promoting Education for Sustainability in the Haredi Sector.*

45. Tal, *The Environment in Israel.*

46. Dunlap, Xiao, and McCright, "Politics and Environment in America," 23–48.

47. Karassin, *True Believers.*

48. For different perspectives on this issue see: Peters, *From Time Immemorial* Morris, *The Birth of the Palestinian Refugee Problem Revisited;* or Masalha, *Expulsion of the Palestinians.*

49. Segev, *1949, The First Israelis*; see also, Israel Central Bureau of Statistics 2011, *Statistical Abstract.*

50. Bassok, "Israel's Population Stands at 7.8 million with 2012 around the Bend," 11.

51. Hann, "Open Spaces in a Urban Society."

52. Israel Central Bureau of Statistics, 2011, *Statistical Abstract.*

53. Udasin, "Rains Are a Little Help to Kinerret Level."

54. Hemu and Bruins, "Climatic Trends in Israel 1970–2002," 63–77.

55. Israel Energy Forum, 2011, NGO website, www.energia.org.il

56. Trilnick, *The Influence of the Landfill Fee on Solid Waste Policy of Local Authorities in Israel.*

57. Tal, "Space Matters."

58. Feitelson, "Malicious Siting or Unrecognized Processes?" 1143–59.

59. Kimchi, "Income Inequality in Israel," 113–52.

60. Feniger and Shavit, "The Demographic Cost Birth Rates and Achievement in International Tests," 303–14.

61. Englard, "Law and Religion in Israel," 185–208.

62. Orenstein, "Population Growth and Environmental Impact," 41–60.

63. Cogan, "Public Expenditure Tables."

64. Cohen and Hareuveini, "The Treasury's Proposal to Reduce Fertility Treatments for Second Child in the Family is Dropped."

65. Rotem, "How Many Children Is Enough for Israelis?"

66. Toledana, Zusman, Farish, Gotlieb, *The Influence of Child Payments on Fertility and Birth Rates.*

67. Tal, *Israel's Environmental Movement,* 121.

68. Ehrlich, Ehrlich, Daily, *The Stork and the Plow, The Equity Answer to the Human Dilemma.*

69. Genesis 1: 28.

70. CIA, "Israel, Economy."

71. Svardilov and Klein, *Masterplan for Treatment of Solid Waste in Israel.*

72. Nisim, *Solid Waste Management in Israel.*

73. Trilnick, *The Influence of the Landfill Fee.*

74. Choshen, "On Foot or Public Transportation?"

75. NET, "Elli Yishai Orders the Potential of Subdividing Existing Apartments."

76. Ben-David, "Sharp Decrease in Water Consumption."

77. Princen, Maniates, and Conca (eds.), *Confronting Consumption.*

78. Tal, *Israel's Environmental Movement,* 23.

79. The Knesset Website: "The Social-Environmental Lobby," http://www.knesset.gov.il/lobby/heb/LobbyPage.asp?lobby=23

80. Red and Green, the Web Site of Knesset Member Dov Henin, http://www.dovblog.org/

81. See a description of the Green Globe event at www.sviva.net

82. Rabinowitz, *Driving Forces.*

83. Richardson and Rootes, eds., *The Green Challenge.*

84. Ehrlich and Holder, "Impact of Population Growth," 1212–17.

85. Tal, "Enduring Technological Optimism," 275–305.
86. Warburg, "Taboo That Needs Breaking."

Bibliography

Amir, S., and O. Rechtman. "The Development of Forest Policy in Israel in the 20th Century: Implications for the Future." *Forest Policy and Economics* 8 (2006): 43.

Avisar, D., Y. Lester., and D. Ronen. "Sulfamethoxazole Contamination of a Deep Phreatic Aquifer." *Science of the Total Environment* 407 (2009): 4278–82.

Barak, A. "The Values of the State of Israel as a Jewish and Democratic State." *Jewish Virtual Library.* 2009. http://www.jewishvirtuallibrary.org/jsource/isdf/text/barak.html

Bassok, M. "Israel's Population Stands at 7.8 million with 2012 around the Bend." *Haaretz.* December 29, 2011.

Ben-David, A. "Sharp Decrease in Water Consumption." *Yedioth Ahronot.* July 2009.

Benstein, J. *The Way into Judaism and the Environment,* 51–53. Woodstock, VT: Jewish Lights, 2006.

Bouskila, A. "Reptiles in Israel." *The Red Book, Vertebrates in Israel.* Tel Aviv: SPNI, 2002. http://www.teva.org.il/?CategoryID=945

Choshen, M. "On Foot or Public Transportation?" *Weblog, Jerusalem* Institute for Israel Studies, February 25, 2011.

CIA. "Israel, Economy." *The World Fact Book,* https://www.cia.gov/library/publications/the-world-factbook/geos/is.html (Accessed January 1, 2012).

Cogan, Y. "Public Expenditure Tables." *State of the Nation Report,* Jerusalem, Taub Center, 2011, p. 102.

Cohen, M., and E. Hareuveini. "The Treasury's Proposal to Reduce Fertility Treatments for Second Child in the Family is Dropped." *Ynet,* December 3, 2003.

de-Shalit, A. "From the Political to the Objective: The Dialectics of Zionism and the Environment." *Environmental Politics* 4, no.1 (1995): 81.

de-Shalit, A., and T. Moti. "Green or Blue and White? Environmental Controversies in Israel." *Environmental Politics* 3, no. 2 (1994): 273–95.

Dunlap, R. E., C. Xiao, and A. M. McCright. Politics and Environment in America: Partisan and Ideological Cleavages in Public Support for Environmentalism." *Environmental Politics* 10, no. 4 (2001): 23–48.

Ehrlich, P., A. Ehrlich, and G. Daily. *The Stork and the Plow, The Equity Answer to the Human Dilemma.* New Haven, CT: Yale University Press, 1997.

Ehrlich, P., and J. P. Holder. "Impact of Population Growth." *Science* 171 (1971): 1212–17.

Englard, I. "Law and Religion in Israel." *The American Journal of Comparative Law* 35, no. 1 (1987): 185–208.

Feitelson, E. "Malicious Siting or Unrecognized Processes? A Spatio-Temporal Analysis of Environmental Conflicts in Tel Aviv." *Urban Studies* 38 (2001): 1143–59.

Feniger, Y., and Y. Shavit. "The Demographic Cost Birth Rates and Achievement in International Tests." *State of the Nation Report,* 303–14. Jerusalem: Taub Center, 2003.

Flash, S. "Long-term Secular Trends in the Age at Menarche in Israel." Presented at Conference on Reproductive Health and Endocrine Disrupting Chemicals. Tel Aviv: Tel Aviv University, November 23, 2009.

Galnoor, I. "Crisis in the Israeli Political System: The Parties as a Predominant Factor." In *Israel toward the Year 2000: Society, Politics, and Culture,* edited by Moshe Lissak and Baruch Knei-Paz, 144–75. Jerusalem: Magnes, 1996 (In Hebrew).

Ginsberg, P. "Afforestation in Israel, A Source of Social Goods and Services." *Journal of Forestry* 98 (2000): 32–36.

Hann, I. "Open Spaces in a Urban Society." In *Israel's Environmental History,* edited by D. Orenstein, A. Tal, C. Miller. Pittsburgh: University of Pittsburgh Press, 2012.

Har-Nir, R., V. Ben-Shoshan, C. Greenfield, I. Eldar, Y. Bdolah, A. Hurwitz, and R. Haimov-Kochman. "Declining Quality of Donated Semen—A 10 Years Historical Comparative Cohort." Unpublished manuscript, May 12, 2009.

Hemu, K., and H. Bruins. "Climatic Trends in Israel 1970–2002: Warmer and Increasing Aridity Inland." *Climatic Change* 96, no. 1–2 (2009): 63–77.

Israel Environmental Protection Ministry. "Ministry Publishes Indicators on the State of the Environment in Israel." News published in ministry web-site: www.sviva.gov.il (2007).

Israel Ministry of Environmental Protection. "Environmental Protection Ministry Publishes Indicators on the State of the Environment in Israel. News published in ministry web-site: www.sviva.gov.il (October 21, 2010).

Kaplan, M., et al. "Patterns in the Utilization of Constructed Land in Israel." Jerusalem: The Jerusalem Institute for Israel Studies. 2007 (In Hebrew, summary in English).

Karassin, O. "The Battle of the 'True Believers': Environmentalism in Israeli Party Politics." In *Israel's Environmental History,* edited by D. Orenstein, A. Tal, and C. Miller. Pittsburgh: University of Pittsburgh Press, (2013), 168–89.

Kimchi, E. "Income Inequality in Israel." *State of the Nation Report—Society, Economy and Policy 2010,* edited by Dan Ben-David, Jerusalem, Taub Center (2011), 113–52.

Lev, O. "Regulated and Non-regulated Microcontaminants in Drinking Water and Wastewater in Israel." Presented at Conference on Reproductive Health and Endocrine Disrupting Chemicals. Tel Aviv: Tel Aviv University, November 23, 2009.

Lubinov, C. *Environmental, Social and Economic Justice and Human Rights.* Tel Aviv: Life and Environment, 2005 (In Hebrew).

Marom-Albeck, O., and A. Tal. "Towards a More Effective Model of Public Interest Environmental Citizen Suits." *Israel Law Review* 34, no. 3 (2000): 373–424.

Matzkin, H. "Testis Cancer in Israel." Conference on Reproductive Health and Endocrine Disrupting Chemicals. Tel Aviv: Tel Aviv University, November 23, 2009.

McKinsey and Company. *Greenhouse Gas Abatement Potential in Israel. Israel's GHG Abatement Cost Curve.* Translated Executive Summary. Herzliyah: Israel, November 2009.

Meizlish, A. *The Battle against the Trans-Israel Highway—Documentation of an Environmental Campaign.* Jerusalem: Jerusalem Institute for Israel Studies, 2005. (In Hebrew).

Morag-Levine, M. "The Politics of Imported Rights: Transplantation and Transformation in an Israeli Environmental Cause Lawyering Organization." In *Cause Lawyering and the State in a Global Era,* edited by A. Sarat and S. Scheingold, 334–53. London: Oxford University Press, 2001.

Morris, B. *The Birth of the Palestinian Refugee Problem Revisited.* Cambridge: Cambridge Middle East Studies, 2004.

Myers, N., R. Mittermeier, C. Mittermeier, A.B. Gustavo, G. da Fonseca, and J. Kent. "Biodiversity Hotspots for Conservation Priorities." *Nature* 403 (2000): 853–58.

Navot, S. *The Constitutional Law of Israel.* Dordrecht: Kluwer, 2007.

Negev, M., Y. Garb, and Jeremy Benstein. *Promoting Education for Sustainability in the Haredi Sector,* unpublished report submitted to the Israel Ministry of Environmental Protection, Jerusalem (2009) (In Hebrew).

Negev, M., G. Sagui, Y. Garb, A. Salzberg, and A. Tal. "Evaluating the Environmental Literacy of Israeli Elementary and High School Students." *The Journal of Environmental Education* 39, no. 2 (2008): 3–20.

Nisim, I. *Solid Waste Management in Israel.* Jerusalem: Israel Ministry of Environmental Protection, Powerpoint presentation, available with author, 2009.

Nur, M. *Expulsion of the Palestinians: The Concept of "Transfer" in Zionist Political Thought, 1882–1948.* Washington, DC: Institute for Palestine Studies, 1992.

Orenstein, D. "Population Growth and Environmental Impact: Ideology and Academic Discourse in Israel." *Population and Environment* 26, no. 1 (2004): 41–60.

Peters, J. *From Time Immemorial: The Origins of the Arab-Jewish Conflict over Palestine.* New York: Harper and Row, 1984.

Preuss, P., Y. Goren, M. Haran, S. Nezer, P. Warburg. "A Comparative Assessment of Air Pollution Public Health Risks in Two Israeli Metropolitan Areas—1995 to 1999." Jerusalem: Israel Ministry of Environmental Protection, 2003.

Princen, P., M. Maniates, and K. Conca, eds. *Confronting Consumption.* Cambridge, MA: MIT Press, 2002.

Public Expert Committee Regarding the State of Air Quality in Israel. *National Masterplan, Clean Air Law,* pp. 4–5. Jerusalem: Ministry of Environmental Protection, March 1, 2011.

Rabinowitz, D., and I. Vardi. *Driving Forces: Trans-Israel Highway and the Privatization of Civil Infrastructures in Israel.* Jerusalem: Van Leer Jerusalem Institute, 2010. (In Hebrew).

Richardson, D., and C. Rootes, eds. *The Green Challenge, The Development of Green Parties in Europe.* London: Routledge, 2005.

Rinat, Z. "Contamination Endangers Coastal Aquifer: One fifth of Israel's Drinking Water." *Haaretz,* February 9, 2007.

Rinat, Z. "Green Groups Warn: Coastal Aquifer Pollution is Limiting Israel's Water Supply." *Haaretz,* March 21, 2011.

Rinat, Z. "International Expert: Israel Needs to Research Its Low Fertility Levels." *Haaretz,* November 24, 2009.

Rotem, T. "How Many Children is Enough for Israelis?" *Haaretz,* July 8, 2011.

Schwartz, E. "Appendix I: Three Paradigms of Environmental Education." In *Paths to Sustainability, Shadow Report to the Government of Israel's Assessment of Progress in Implementing Agenda 21,* p. 38, presented at The World Summit for Sustainable Development. Johannesburg, Tel Aviv: Heschel Center for Environmental Learning and Leadership, 2002.

Segev, T. *1949, The First Israelis.* New York: The Free Press, 1986.

Shalmon, B. "Mammals in Israel." *The Red Book, Vertebrates in Israel.* Tel Aviv: SPNI, 2002, http://www.teva.org.il/?CategoryID=949

Shenhav, Y. *What is Multiculturalism?* Tel Aviv: Babel, 2005. (In Hebrew).

Sosner, Y. *The Environmental Campaign as a Force for Molding a New Identity, The Katamon Neighborhood's Campaign Against Building in the Pre-Har Valley.* PhD diss., Tel Aviv University, 2009.

Svardilov, U. Erez, and D. Klein. *Masterplan for Treatment of Solid Waste in Israel.* Jerusalem: Israel Ministry of Environmental Protection, Jerusalem, 2005.

Tal, A. "Enduring Technological Optimism: Zionism's Environmental Ethic and its Influence on Israel's Environmental History." *The Journal of Environmental History* 13(2008): 275–305.

Tal, A. *The Environment in Israel: Natural Resources, Crises, Campaigns and Policy from the Advent of Zionism until Twenty-first Century.* B'nei Brak: HaKibbutz HaMeuhad Press, 2006. (In Hebrew).

Tal, A. *Pollution in a Promised Land—An Environmental History of Israel.* Berkeley: University of California Press, 2002.

Tal, A. "Seeking Sustainability: Israel's Evolving Water Management Strategy." *Science* 313 (2006):1081–84.

Tal, A. "Space Matters: Historic Drivers and Turning Points in Israel's Open Space Protection Policy." *Journal of Israel Studies* 13, no. 1 (2008): 119–51.

Tal, A., A. Ben-Gal, P. Lawhon, and D. Rassass. *Sustainable Water Management in the Drylands: Recent Israeli Experience.* Jerusalem: Israel Ministry of Foreign Affairs, October 2005.

Tal, A., S. Leon-Zchut, L. Frankel-Oshri, I. Greenspan, and S. Akov. *Israel's Environmental Movement: Trends, Needs and Potential.* Submitted to the chairman of the Knesset and the Knesset Environmental Committee. Beer Sheva: Ben Gurion University, June21, 2011.

Tamir, T. "Two Concepts of Multiculturalism." *Multiculturalism in A Democratic and Jewish State,* edited by M. Moutner, A. Sagy and R. Shamir, 79–92. Tel Aviv: Ramot Yona, 1998.

Toledana, E., N. Zusman, R. Farish, and R. Gotlieb. *The Influence of Child Payments on Fertility and Birth Rates.* Daniel, Research Paper 101. Jerusalem: Central Bureau of Statistics, 2009.

Trilnick, I. *The Influence of the Landfill Fee on Solid Waste Policy of Local Authorities in Israel.* Master's thesis, Ben Gurion University, 2012.

Udasin, S. "Rains Are a Little Help to Kinerret Level." *Jerusalem Post,* November 18, 2011.

Warburg, P. "A Taboo That Needs Breaking." *Jerusalem Post*, January 1, 1997.
Waskow, A., ed. *Torah of the Earth, Exploring 4,000 Years of Ecology in Jewish Thought*, vols.1 and 2. Woodstock, VT: Jewish Lights, 2006.
YNET. "Elli Yishai Orders the Potential of Subdividing Existing Apartments." *Ynet*, internet newspaper, June 11, 2011.

Conclusion: Environmentalism and Climate Change: Green Politics in an "Age of Stupid"

Sherilyn MacGregor and Timothy Doyle

"Why didn't we stop climate change when we had the chance?" This is the provocative question that frames *The Age of Stupid,* a British eco-documentary released in 2009. It is a film offering a dystopian portrait of a devastated world in 2055 and a backward look at missed opportunities for averting the "suicide" of the species. It is a cautionary tale about the price that future generations will pay for the ignorance and stupidity (i.e., the greed, the oil addiction, the NIMBYism, the air travel, etc.) of the current generation. To tell the tale, the keeper of a global archive and the last man on earth shows us a montage of video footage from the lives of six real people living in 2008, in different parts of the world, who are all in different ways affected by or implicated in environmental destruction. The archivist wonders aloud: "what state of mind were we in, to face extinction and simply shrug it off?" The message is powerful and frightening, and the film has been widely celebrated for providing a wake-up call that is more radical and hip than Al Gore's Nobel Prize–winning Powerpoint presentation in *The Inconvenient Truth.*

Although some call the film depressing and sensationalist, it has spawned the rapidly growing environmental campaign called "10:10" in the United Kingdom. Organized by the film's director Franny Armstrong, the 10:10 campaign is a project that aims to bring all sectors of British society together to agree to cut their emissions by 10 percent by the end of 2010. Tens of thousands of individuals and organizations have signed up to take the 10:10 pledge. The campaign website provides the rationale that "politicians have so far failed to do what needs to be done, so it's time for ordinary people to step in and show that we're ready to defend our children's futures. It's now or never for the climate." Sounds like a call to a reinvigorated, global environmentalism with new strategies and added urgency. Or does it? A closer look gives reason to despair. One need only visit the website to find a familiar list (which can be downloaded and put on the fridge) of 10 "simple ways" to cut carbon, a list that looks no different from the individualized, private acts of green consumerism and energy conservation (e.g., "drive less, eat better, don't waste water". . .) that the environmental movement has been espousing since the 1970s. This time it is slick and well marketed, with celebrity endorsements and a piece of jewelry to (buy from the website and) wear as an outward statement of having taken the pledge. But nowhere to be found: the words "environmentalist" or "citizen."

* * * * * *

Global environmental problems such as climate change are often used as reasons for why environmentalisms can no longer be confined to the traditional territorial spaces of nation states. As Tim Doyle and Sanjay Chaturvedi observe, "In both popular and scientific discourses, climate change emerges as the paradigmatic global environmental 'threat beyond borders' in need of global consequence, action and accords."[1] Climate change is arguably one of the most significant drivers of recent academic interest in global environmental governance and the search for some form of green cosmopolitanism. From civil society there are also loud calls for global cooperation, for the leaders of the world to work together beyond borders and national interests for the sake of common humanity and planetary survival. Climate change has also become a dominant concern within nation states; it has for several years shaped the environmental agenda in the U.K. context in which we write. As a nation with a long tradition of environmentalism, it is not surprising that the climate change issue would take root so deeply in the past five years or so. The most recent wave of interest arguably was sparked by the 2004 warning by the chief scientist that it is a threat worse than terrorism.[2] Then, in 2008, after several years in the making, the government passed the Climate Change Act, making the United Kingdom the first country in the world to have a legally binding, long-term framework to cut carbon emissions. Alongside these government actions, a political movement has coalesced around climate change. All environmental organizations have put aside their specialist causes (e.g., birds, woodlands, transport) to join the climate cause. For example, the Stop Climate Chaos Coalition includes over 100 different organizations and claims to have 11 million members. On the other side of the issue, there is a small group of "climate skeptics" whose opposition to climate change policy and science is widely regarded in the media to be funded by corporate interests and hence generally has been dismissed by the mainstream pro-environmental public. The dismissal of climate skepticism as a fringe element may be showing signs of change, however, as in the past year a significant proportion of the U.K. population is starting to express doubts. A BBC opinion poll recently found that the percentage of people who believe that climate change is largely caused by humans fell from 50 percent to 34 percent between late 2009 and early 2010.[3] Many commentators believe that this rise in skepticism may have been influenced by the economic recession, or by a very public scandal of (allegedly) dodgy data handling in the Climate Research Unit at the University of East Anglia (called "Climategate"), or by the particularly bad weather in the United Kingdom in the past four years.

The recent spate of climate skepticism notwithstanding, it is fair to say that the high level of public concern about climate change that has been apparent in the United Kingdom and many other countries around the world gives many people hope. Many wish to focus on the activism, the coalition politics, the global dialogue, and transnational cooperation. Could this be the beginning of a reinvigorated, counter-hegemonic, global environmentalism based on cosmopolitan values and a rejection of ecocidal neoliberalism? We think not. There are several aspects of climate change politics that move in an opposite direction to the kind of environmentalism that we wish to envision. There are in fact several dimensions of a dominant climate-centric environmentalism in the majority world that may actually serve to support the continuation of neoliberalization while presenting serious challenges to the goals of democracy, justice, and environmental change both within countries and globally. We base this case on the following three observations.

First, despite the scientific evidence to suggest the de-territorializing nature of climate change, emerging geopolitical as well as geo-economic discourses on climate change tend to re-territorialize a range of political issues. For example, debates over what is a "fair" distribution of greenhouse gas reductions always operate at the level of national measurements (carbon footprints) and the rearticulation of the global carbon cycle as carbon sinks is the result of a process whereby the flows of carbon are shaped to coincide with the borders of the sovereign state.[4] International relations scholars have also pointed to the rise of a climate security agenda that presents climate change as a threat to national security and sovereignty. They observe the emergence of a neo-Hobbesian fear that climate-related resource scarcity will lead inevitably to political instability within, and conflict between, nations as well as growing trends toward policing climate change. This situation in turn may lead to militarized approaches and a renewed importance of border controls. Doyle and Chaturvedi argue that mass migrations of people from the global South are constructed as a risk, and something the North needs protecting against.[5] Indeed concerns about so-called climate refugees moving to more desirable places (i.e., Europe) may signal a reassertion of the strong borders that globalization was said to have rendered less significant.

While proponents of cosmopolitanism have a complex and often contradictory approach to territory and the nation state (often being both antinationalist and believers in the need for institutions of the nation state to deliver goods), they tend to hold universal human rights and "kindness to strangers" as core moral principles.[6] Rather than being seen as citizens or partners in the global effort to combat climate change, however, people in the global South are increasingly portrayed as a threat, either as potential refugees wanting to invade "our shores" or as rapidly multiplying aspirants of a Western way of life. The clear villain of the piece in *The Age of Stupid* is Jeh Wadia, the wealthy Indian founder of the budget airline GoAir, who wants to offer one-rupee flights so that everyone in India ("even rickshaw drivers and servants") can travel with the same ease as Europeans. Viewers are encouraged to fear his ambition when we see news reports about "the worst rainfall in living memory in Mumbai" and hear others saying that the carbon emitted by air travel "causes the death of people." We would suggest that the territorialized climate security discourse, coupled with xenophobia and racist fear of the Other, runs counter to open, intercultural dialogue and may even signal a rescinding of the concept of hospitality that is so central to cosmopolitanism.[7] It may also be reasonable to worry that in times of crisis, exceptional measures for curbing/controlling/monitoring the environmentally destructive behavior of (stupid) citizens may be seen as legitimate, where the restraints of the regular political sphere, including human rights, can be set aside in the name of national/climate security. The rise of neo-Malthusian discourse, which is almost exclusively interested in population growth in the global South, and which suggests that there might be cause for considering coercive measures as a last resort in order to tackle climate change, is particularly worrying in this regard.

Second, against those who frame climate change as a global issue that all the world's citizens can and should get behind, it is possible to see climate politics as highly exclusive and exclusionary. Its definition as a problem and the ways in which solutions are being devised rests in the hands of an elite group of mainly male, mainly white scientists, policymakers and capitalists from the affluent Western world. Its identification as *the most urgent environmental problem of our time* has resulted in the homogenization of what was once a diverse environmental movement. It has excluded less powerful people (the majority world) from the debate and has sidelined a range of other environmental issues that are not immediately about climate or carbon. Eileen Crist argues that this has resulted in "the restriction of proposed solutions to the technical realm . . . and it detracts attention from the planet's ecological predicament as a whole, by virtue of claiming the limelight for the one issue that trumps all others."[8]

An important postcolonial critique of the domination by climate change of the global environmental agenda is that it not only reflects minority world experiences and interests but also that it serves as an instrument for maintaining class power and imperial domination. This echoes the position taken by critics of the 1990s sustainable development agenda and of the "globalitarian" ethics expressed by the Earth Charter.[9] Heather Smith argues that climate change politics consistently marginalizes indigenous peoples and that its definition "as global masks alternative voices that fundamentally challenge Western ways of knowing, being and doing."[10] Sherilyn MacGregor has recently written that due to the domination of elite male scientists and policy-makers at all levels, climate change has brought about the masculinization of environmentalism.[11] One result of climate change becoming the dominant issue is that most other environmental issues—those especially relevant to poor people and women—have been sidelined, even though they have not been solved.

The fact that climate change mitigation and adaption strategies tend to be technical (like nuclear power and genetically modified organisms (GMOs)) not only reflects a stereotypically masculinist approach to problem solving, but is also very good for neoliberal business. The discourse of "ecological modernization," now dominant in Europe, calls for the cooperation between government and business to solve environmental problems—in a win-win-win situation. Climate change represents a welcome opportunity for this worldview by providing a serious global problem that can be solved by the wise partnering of techno-innovators and brave capitalists. International carbon trading, geo-engineering, and clean technology transfer from North to South promise to make a lot of profit for transnational capitalists (some are calling it carbon capitalism). These techno-fixes are not aimed at addressing the needs of the majority of people living in poverty (70 percent of whom are women) for whom daily survival is more pressing than energy-saving gadgets. It is highly likely that no one has ever asked them what their needs are for climate change adaptation or mitigation. From the perspective of those marginalized by climate politics (most of whom bear comparatively little responsibility for causing the earth's climate to change), an obvious response to *The Age of Stupid* question "Why didn't we stop climate change when we had the chance?" is: "Who is we?" As Marx once said, sometimes the only answer to a problematic question is to question the question.[12]

Third, to the extent that climate change politics invites individualized responses from citizens, it is depoliticizing environmentalism. So-called "climate action" tends to involve lifestyle choices rather than citizenly acts, and people are signing up to pledge their personal commitment in droves. It is symptomatic of the triumph of the ultimate neoliberal subject—the citizen-consumer—that people have internalized personal responsibility for the problem along with the idea that the best way to respond is in the private, not public, sphere. Individualism is a core idea of neoliberalism; it is not being challenged by the politics of climate change. Not only is the word "citizen" typically absent from the discourse, but governments and Western environmentalists place emphasis on the role of individuals *as consumers* to tackle climate change by conserving energy, taking public transit, recycling waste, growing food, and foregoing flights. For example the 10:10 campaign, which has apparently given up on politicians to lead the charge, asks individuals to adopt an almost identical list of behaviors in their everyday lives, the majority of which either involve consumption choices or work in the private household. Individuals are presented in gender- and class-neutral terms, as if all share the same ability to make the necessary changes and the same burden of work that these changes will entail. This privatization and individualization of carbon reduction duty is obviously problematic from a feminist perspective. Gender-blind policies and strategies carry the risk of perpetuating existing inequalities.

Environmental campaigns that focus on the duty of the individual citizen-consumer are not new in the postmaterialist global North. In a 2001 article titled "Plant a tree, buy a bike, save the world?" Michael Maniates discusses the problem of individualization that is symptomatic of environmentalism's "flight from politics" over the past 20 years.[13] He suggests that the desire to avoid painful, confrontational political struggle has combined with the internalization of the neoliberal rhetoric of consumer choice to produce an increasingly ineffectual movement. Not specifically addressing global warming at the time, he surmises that the complexity and increasingly global scale of environmental problems may have something to do with this depoliticization of environmentalism. 10 years later, in 2010, climate activists in the United Kingdom and the United States adopt the same individualist discourse with added urgency—and it is now widely used by governments, businesses, NGOs, and eco-activist organizations alike. Are we all environmentalists now?

Examples of individualistic low-carbon lifestyle strategies are rapidly proliferating and yet are more often than not accessible only to the affluent. The heroes of *The Age of Stupid* are a painfully middle-class family of five from Cornwall who have a waste-oil powered car, own their own wind turbine (Dad is a wind farm developer), have enough land that they are almost self-sufficient in food and, after much difficult deliberation, decide to turn down a skiing trip in France (because, Mum says, "after setting fire to a forest, flying is the single worst thing one can do to cause climate change!"). From carbon offsetting to planting vegetable gardens (all very expensive endeavors, it is important to note), people like them are increasingly equating the fight against climate change with personal choice and/or sacrifice. This is surely a problem for environmentalism: it is in danger of losing its status as an oppositional social movement. Where there are only individual acts and no collective stance against the dominant order, there is no political movement. Maniates puts it nicely when he argues that "when responsibility for environmental problems is individualized, there is little room to ponder institutions, the nature and exercise of political power, or ways of collectively changing the distribution of power and influence in society—to, in other words, 'think institutionally.'"[14] What will happen if there is no longer a counter-hegemonic environmental movement, but only one big global consensus about saving the planet one energy-saving light bulb at a time?

Finally, the preceding points combine to make us suspect that the politics of climate change is a threat to the project of reinvigorating environmentalism as part of a practice of oppositional citizenship in a green public sphere. We have several short points about this fourth problem—each deserves more interrogation and illustration than we have space for here. To begin with, it is evident that climate politics either ignores or instrumentalizes citizenship. The version of green citizenship that is currently being promoted as a response to the climate crisis by governments and environmentalists alike commonly relies on an unproblematized conception of the environment (or the climate) as a domain known by science, and grants citizenship a correspondingly unproblematized role *in service to it.* To the extent that it even enters the discourse, citizenship is instrumentalized as a means to a planet-saving end. There is little or no acknowledgment that there might be other values of citizenship or that the practice of citizenship might be valuable in and of itself. Without this, what chance is there for reestablishing the line between consumer and citizen, or for defending citizenship as a specific political identity that is fundamentally not the same as consumership and that is necessary for human flourishing?

With climate change, there is little room for politics when an inaccessible, authoritative science sets the terms and the parameters of the issue. Here we are reminded of Michel Foucault's rejection of any kind of politics based on foundational or universal principles. He argues that moral universals suggest the existence of Truth, which in modern times is simply bourgeois ideology dressed up as science. Similarly, Hannah Arendt also refers to science as a private, depoliticized kind of knowledge that leaves no room for debate. Applying these observations to environmentalism, Douglas Torgerson expresses disagreement with the way greens tend to use "facts" about the impending destruction of the environment to support a program for change that needs (urgent) implementation rather than ongoing, democratic negotiation. "Politics" he says, is a "juggling act that defies fixed concepts."[15] The understanding of citizenship that we take from this point is that it is inherently contested and self-reflexive. So what does it mean when citizens can neither reflect on their role as citizens nor engage in political debate in the public sphere? It must be a bad thing for citizenship if people are not able to disagree with or challenge the way climate politics is evolving without being labeled a climate denier—or "stupid."

* * * * * *

The chapters in this collection go some way to reminding us that in spite of the domination of the environmental agenda by the all-encompassing climate story, there are many environmental movements around the world that either resist or are almost oblivious to this framing. Our collection includes 24 countries from six major regions of the world: South America, North America, Africa and the Middle East, Europe, Asia, and Oceania. With the exception of the chapters on countries in North America, Western Europe, Australia and New Zealand—the global North—most of our collection offers empirical evidence that there is more to the environmental story than climate. Globalization—of culture and academic information—often blinds us to the diversity of interpretations that actually exist in the world. A motivation for this collection is our belief that one of the most important ways that those of us who study green politics can participate in moving toward oppositional-cosmopolitan environmentalisms is through empirical research. We must try not to theorize about global problems only from our desks, nor should we make assumptions about or speak on behalf of those who inhabit the margins through no choice of their own. We need more scholarship on alternative, local enactments of popular environmental struggle and to give voice to people who have been marginalized by the global climate story.

Shortly before the end of *The Age of Stupid,* the archivist says that the only way we could have stopped the suicide of the species was by "questioning our collective values." Ironically, by using the stories of six different people in six different places (the United States, Jordan, the United Kingdom, Nigeria, India, and France) to make the case for a shared crisis of humanity, the film demonstrates that there are in fact very wide disparities in values and in how people experience environmental problems in daily life. The resistance of the Ogoni people against Shell's occupation of the Niger Delta is rooted in a different set of values from the resistance of French cyclists against truck traffic on the motorways. For us, this was the most important message in the film: local stories need to be told, understood, and compared before we try to imagine what a more cosmopolitan environmentalism might be or might be for. We need to root ourselves in particular places and communities so we can listen to the stories of the people who are frequently presented as numbers on a population graph or as one-dimensional objects (victims, villains, climate changers) in the climate discourse. Gathering local examples of resistance might contribute to a more complete picture of the global situation that can help us to challenge the dominance of monolithic, neoliberal, Eurocentric, masculinist climate politics. There are many voices, neither green nor stupid, that deserve to be heard and many questions to be asked. We academics of the minority world have a responsibility to use our privilege to make it happen.

Notes

1. Doyle and Chaturvedi, "Climate Territories: A Global Soul for the Global South?" 517.
2. Connor, "U.S. Climate Policy Bigger Threat to World than Terrorism."
3. Corbyn, "Climategate: The Cold Comfort Is That It Is Now a Hot Topic."

4. Paterson and Stripple, "Singing Climate Change in to Existence: On the Territorialisation of Climate Policy-making."

5. Doyle and Chaturvedi, "Climate Territories: A Global Soul for the Global South?"

6. Appiah, *Cosmopolitanism: Ethics in a World of Strangers.*

7. See Held, "Globalisation, Corporate Practice and Cosmopolitan Social Practice."

8. Crist, "Beyond the Climate Crisis: A Critique of Climate Change Discourse," 32.

9. MacGregor, "Reading the Earth Charter: Cosmopolitan Environmentalism or Light Green Politics as Usual?"

10. Smith, "Disrupting the Global Discourse of Climate Change: The Case of Indigenous Voices," 198.

11. MacGregor, "A Stranger Silence Still: The Need for Feminist Social Research on Climate Change."

12. Harvey, *Cosmopolitanism and the Geographies of Freedom,* 260.

13. Maniates, "Individualization: Plant a Tree, Buy a Bike, Save the World?"

14. Ibid., 33.

15. Torgerson, *The Promise of Green Politics: Environmentalism and the Public Sphere,* 103.

Bibliography

Appiah, A. K. *Cosmopolitanism: Ethics in a World of Strangers.* London: Penguin, 2006.

Arendt, H. *The Human Condition.* Chicago: University of Chicago Press, 1958.

Connor, S. "US Climate Policy Bigger Threat to World than Terrorism." *The Independent,* January 9, 2004. http://www.independent.co.uk/news/world/americas/us-climate-policy-bigger-threat-to-world-than-terrorism-572493.html (Accessed May 24, 2007).

Corbyn, C. "Climategate: The Cold Comfort Is That It Is Now a Hot Topic." *Times Higher Education* 8 (2010): 10–11.

Crist, E. "Beyond the Climate Crisis: A Critique of Climate Change Discourse." *Telos* 141 (Winter 2007): 29–55.

Doyle, T., and S. Chaturvedi. "Climate Territories: A Global Soul for the Global South?" *Geopolitics* 15 (2010): 516–35.

Harvey, D. *Cosmopolitanism and the Geographies of Freedom.* New York: Columbia University Press, 2009.

Held, D. "Globalisation, Corporate Practice and Cosmopolitan Social Practice." *Contemporary Political Theory* 1, no.1 (2002): 59–78.

MacGregor, S. "Reading the Earth Charter: Cosmopolitan Environmentalism or Light Green Politics as Usual?" *Ethics, Place and Environment* 7, no. 1–2 (March/June 2004): 85–96.

MacGregor, S. "A Stranger Silence Still: The Need for Feminist Social Research on Climate Change." *Sociological Review* 57 (2009): 124–40.

Maniates, M. "Individualization: Plant a Tree, Buy a Bike, Save the World?" *Global Environmental Politics* 1, no. 3 (2001): 31–52.

Paterson, M., and J. Stripple. "Singing Climate Change in to Existence: On the Territorialisation of Climate Policy-making." In *The Social Construction of Climate Change: Power, Knowledge, Norms, Discourses,* edited by M. Pettinger, 149–72. Aldershot: Ashgate, 2007.

Smith, H. "Disrupting the Global Discourse of Climate Change: The Case of Indigenous Voices." In *The Social Construction of Climate Change: Power, Knowledge, Norms, Discourses,* edited by M. Pettinger, 197–215. Aldershot: Ashgate, 2007.

Torgerson, D. *The Promise of Green Politics: Environmentalism and the Public Sphere.* Durham: Duke University Press, 1999.

Index

Page numbers in *italics* indicate tables.

About the Editors and Contributors

Editors

TIMOTHY DOYLE, BA Hons (Melb), MA (Adel), PhD (Griffith), is professor of politics and international studies at the University of Adelaide in Australia, and founding chair of management for the Indo-Pacific Governance Research Centre (IPGRC), also at Adelaide. Professor Doyle is also chair of politics and international relations at Keele University in the United Kingdom. He has published widely in a diverse range of journals. His most recent works include the books: *Environmental Politics and Policy-Making in Australia*; *Environment and Politics* with D. McEachern; *Green Power: The Environment Movement in Australia*; *Environmental Movements in Minority and Majority Worlds: A Global Perspective*; *Beyond Borders: Environmental Movements and Transnational Politics*, edited with B. Doherty; and *Crucible for Survival: Environmental Security and Justice in the Indian Ocean Region*, edited with M. Risely. He has taught and contributed to university courses in the United Kingdom, the United States, Malaysia, India, and Australia and been a dedicated environmental and human rights activist since the 1980s. He serves on a variety of editorial boards and has a long and extensive background in editing and co-editing.

SHERILYN MACGREGOR is senior lecturer in environmental politics at Keele University in the United Kingdom. Her published works include *Beyond Mothering Earth: Ecological Citizenship and the Politics of Care* (University of British Columbia Press, 2006) and numerous journal articles and book chapters on gender, feminist theory, and environmental justice, including "A Stranger Silence Still: The Need for Feminist Social Research on Climate Change" (*Sociological Review* 2009). MacGregor is joint editor of

Environmental Politics journal and a convener of the Environmental Politics Standing Group of the European Consortium for Political Research. She holds a PhD in environmental studies from York University in Toronto, Canada.

Contributors

LAURIE ADKIN, PhD, is associate professor of political science at the University of Alberta, Edmonton, AB. Her published works include *Politics of Sustainable Development: Citizens, Unions, and the Corporations*; *Environmental Conflict and Democracy in Canada*; and essays on social movements theory, political ecology, and family policy. She holds a doctorate in political studies from Queen's University in Kingston, ON.

JOHN AGBONIFO, PhD, is assistant professor of sociology at the Osun State University, Osogbo, Osun State. His published works include "Development as Conflict: Ogoni Movement, the State and Oil Resources in the Niger Delta," "Nigeria: Oil, Insecurity, and Subversive Patriots in the Niger Delta: The Ogoni as Agent of Revolutionary Change"; "Corporate Social Responsibility: An Oversocialized View of Multinational Corporations in Africa?"; "Territorialising Niger Delta Conflicts: Place and Contentious Mobilisation"; and "Environmentalism and Environmental Movements in Nigeria." John holds a doctorate in development studies from the International Institute of Social Studies, Erasmus University.

MAURICIO BERGER, PhD, is postdoctoral fellow at the National Council of Scientific and Technological Research of Argentina (CONICET), and professor at the Universidad Nacional de Córdoba, Argentina. His published works include the co-edition of the Latin American issue of the *Environmental Justice* journal (April 2012), and the co-edition and co-authorship of books about environmental justice and citizenship in Argentina: *Justicia Ambiental y Creatividad Democrática* (2012), *Justicia Ambiental. Saberes prácticos para la efectiva vigencia de los derechos ambientales,* and *Estado incivil y ciudadanos sin estado,* and *Estado incivil y ciudadanos sin estado* (2009). Mauricio holds a doctorate in social sciences from the Universidad de Buenos Aires, Argentina.

ROBERT BRULLE is a professor of sociology and environmental science in the department of culture and communications, and an affiliate associate professor of public health in the School of Public Health at Drexel

University in Philadelphia, Pennsylvania. He is the author of over 50 articles on the U.S. environmental movement, critical theory, and public participation in environmental policy-making, and is the author of *Agency, Democracy and the Environment: The U.S. Environmental Movement from the Perspective of Critical Theory*, as well as co-editor, with David Pellow, of *Power, Justice and the Environment*. He has a BS degree in marine engineering from the U.S. Coast Guard Academy, an MA in sociology from the New School for Social Research, an MS in natural resources from the University of Michigan, and a PhD in sociology from George Washington University.

BENITO CAO, PhD, is lecturer in politics at the University of Adelaide, Australia. His published works include essays in the fields of pedagogy, political culture and Brazilian national identity. He is currently working on a manuscript for the Routledge *Introductions to Environment* series entitled *Environment and Citizenship*. He holds a doctorate in politics from the University of Adelaide.

MARÍA PILAR GARCÍA-GUADILLA, PhD, is professor of sociology and political ecology at the Universidad Simón Bolívar, Caracas, Venezuela. Her published works include *Challenging Development and Democracy: The Environmental Social Movements in Europe and Latin America*; *Ambiente, Estado y Sociedad: Crisis y Conflictos socio-ambientales en Venezuela y América Latina*, and essays on environmental conflicts, strategic planning, and social movements. García-Guadilla holds a doctorate in urban sociology and human ecology from the University of Chicago and a post-doctorate in social movements from the University of London.

JANET MUTHONI MUTHUKI, is a lecturer of gender studies at the University of KwaZulu-Natal, Durban, South Africa. Her published works include: "Challenging Patriarchal Structures: Wangari Maathai and the Green Belt Movement in Kenya" (in *Agenda*); "The Politics of Relocation and the Negotiation of Family Relationships across Transnational Space" (in *Alternation: Journal of the Centre of Southern African Literature and Languages*); and "Gender Dynamics in Qualitative Research: A Personal Reflection" (in *Ikamva—Journal of Social Sciences and Humanities*). She holds a doctorate in gender identity (re)construction and transnationalism from the University of KwaZulu-Natal.

M. A. MOHAMED SALIH, PhD, is professor of politics of development both at the International Institute of Social Studies, Erasmus University,

Rotterdam and the department of political science, University of Leiden. His published works include *Local Climate Change and Society* (Routledge); *Climate Change and Sustainable Development Challenges for Poverty Reduction in the 21st Century* (Edward Elgar); and *Interpreting Islamic Political Parties* (Palgrave).

CLAUDIA SEPÚLVEDA, PhD, is associate researcher at the Center of Environmental Studies, Universidad Austral de Chile, Valdivia, Chile (Centro Transdisciplinario de Estudios Ambientales y Desarrollo Humano Sostenible (CEAM), Universidad Austral de Chile.). Her published works include two books and several chapters and papers on environmental conflicts in Chile, mainly oriented to policy recommendations. Sepúlveda holds a master's in human settlements and environment and her doctoral research in human geography, at the University of British Columbia, is aimed to trace the long-lasting institutional effects of the Río Cruces disaster in Valdivia from a post-humanist perspective.

ALON TAL is a researcher at Ben Gurion University of the Negev's Desert Research Institutes. Prior to this post, he was the founding director of the Israel Union for Environmental Defense, the country's leading environmental advocacy organization, and the Arava Institute for Environmental Studies, an advanced training center for Middle Eastern students. Today he is also chair of Israel's green party. His previous books include *Pollution in a Promised Land*; *Speaking of Earth*; *Water Wisdom*; *Building Green, Promoting Energy Efficiency in Israel*; and *All the Trees of the Forests.*

PABLO VILLARROEL, PhD, is associate professor of science and environment at the Social Communication Institute, Universidad Austral de Chile, Valdivia, Chile (Instituto de Comunicación Social, Universidad Austral de Chile). His published works include topics on science, communication, and environmental policies. Villarroel holds a master's in human settlements and environment and is pursuing his doctorate research in reflexivity within environmental conflicts at the Universidad Autónoma de Barcelona, Spain.